More Praise for *Saving the Big Sky*

"*Saving the Big Sky* is a fine blend of history, culture, and land conservation success stories. The beautiful illustrations complement the authors' inspiring accounts of how a mix of private landowners, tribes, land trusts, and funders have conserved more than six million acres of Montana's landscapes." —TOM DANIELS, author of *The Environmental Planning Handbook: For Sustainable Communities and Regions*

"This is an extraordinarily well researched, in-depth history of what led to the creation of the many conservation easements now in place around the state of Montana, and I know it will be referred to for many decades. I found it to be an absolutely fascinating read."
—LAND M. LINDBERGH, rancher and conservationist

"Montana is robust and precious. Hereabouts we treasure what's natural and wild. At a time when so many Americans get their notions of the state from a cheesy horse opera titled *Yellowstone*, it couldn't be more important to present certain realities—about what's worth conserving, and how best to conserve that—in a tapestry made vivid not by fantasy but by fact. Bugbee, Kiesling, and Wright do that superbly (with a bit of crucial help from Shane Doyle) in this deeply informed, loving book." —DAVID QUAMMEN, *New York Times* bestselling author of *The Song of the Dodo: Island Biogeography in an Age of Extinctions*

"The maps! They light you right up with hope. In each chapter focused on a key Montana region, you'll discover a pair of maps contrasting the acreage safeguarded as of 1970 with the total in 2021. This is what a half-century of galloping conservation success looks like: Protected areas proliferate in number and size. More link to others. Whole countrysides—whole ecosystems—become connected through preserved public lands, tribal lands, and impressive arrays of private lands where owners have arranged conservation easements on their property. What's driving all this? The book offers plenty of case histories and compelling perspectives to help answer the question. Beyond that, well. . . . I reckon most everybody around here just knows that a big sky looks best over unspoiled landscapes."
—DOUGLAS H. CHADWICK, author of *The Wolverine Way*

OREGON STATE UNIVERSITY PRESS CORVALLIS

Saving the Big Sky

A Chronicle Of Land Conservation In Montana

Bruce A. Bugbee
Robert J. Kiesling
John B. Wright

With contributions from the Confederated Salish and Kootenai Tribes, Shane Doyle (Apsáalooke), Steve Running, and Todd Wilkinson
Maps supervised by Kevin McManigal and Hannah Shafer
Photographs by Kevin League

Oregon State University Press in Corvallis, Oregon, is located within the traditional homelands of the Mary's River or Ampinefu Band of Kalapuya. Following the Willamette Valley Treaty of 1855, Kalapuya people were forcibly removed to reservations in Western Oregon. Today, living descendants of these people are a part of the Confederated Tribes of Grand Ronde Community of Oregon (grandronde.org) and the Confederated Tribes of the Siletz Indians (ctsi.nsn.us).

Library of Congress Cataloging-in-Publication Data

Names: Bugbee, Bruce A. author | Kiesling, Robert J. author | Wright, John B. (John Burghardt), 1950- author
Title: Saving the Big Sky : a chronicle of land conservation in Montana / by Bruce A. Bugbee, Robert J. Kiesling, John B. Wright ; with contributions from the Confederated Salish and Kootenai Tribes, Shane Doyle (Apsáalooke), Steve Running, and Todd Wilkinson ; maps supervised by Kevin McManigal and Hannah Shafer ; photographs by Kevin League.
Description: Corvallis, OR : Oregon State University Press, 2025. | Includes bibliographical references and index.
Identifiers: LCCN 2025001405 | ISBN 9781962645362 hardcover | ISBN 9781962645379 ebook
Subjects: LCSH: Natural areas—Montana | Protected areas—Montana | Land use, Rural—Montana | Nature conservation—Montana
Classification: LCC QH76.5.07 B84 2025 | DDC 333.7209786—dc23/eng/20250303
LC record available at https://lccn.loc.gov/2025001405

∞ This paper meets the requirements of ANSI/NISO Z39.48-1992 (Permanence of Paper).

Designed by Erin Kirk

First published in 2025 by Oregon State University Press
Printed in China

121 The Valley Library
Corvallis OR 97331-4501
541-737-3166 • fax 541-737-3170
www.osupress.oregonstate.edu

DEDICATIONS

Bruce Bugbee: I have been fortunate to have fine mentors who care deeply about our world, especially my partners in creating this book, Jack and Bob, and most of all my loving partner in life, Nancy.

Bob Kiesling: To all the Montanans who have worked so diligently over the past half century in elevating conservation to its rightful place as a major public and private good, and especially to my colleagues in this endeavor, Bruce and Jack.

Jack Wright: To Bruce and Bob, 50-year friends and coworkers in conservation. To Montana landowners who stood up for the good Earth. To my beloved wife, Rachel; my dear sister Cathy; and Micah—the best poodle in the world.

CONTENTS

ACKNOWLEDGMENTS

The list of people to thank is as long as Montana is wide. So many people helped us create this version of a shared story of land conservation. Special thanks go to Helena Maclay, who guided us in establishing a partnership with the Foundation for Montana History. At the foundation, Charlene Porsild, Ciara Ryan, and Jacob Rosen helped us immensely. Kim Hogeland of Oregon State University Press was our patient mentor in seeing this book through to publication. Thanks to Laurel Anderton for her excellent and thorough copyediting. Deep appreciation to Bill Wyckoff, Drew Bennett, and Tom Daniels for reviewing the text and providing helpful critique. Of course, our deep gratitude goes to Kim Hogeland, Katherine White, Micki Reaman, and everyone at Oregon State University Press for getting this book in print.

We thank all who donated to fund this expensive project: Michelle Bazzanella, Robert Stephens Endowment, Mary Sexton, John C. Schmidt, Randy Gray, Bill and Peggy Strong, Stan and Glenda Bradshaw, Bill and Pam Bryan, Donn J. McAfee, Bitterroot River Lodge, Cheryl S. Garrett, Pennie Strong Tague and Gary Tague, Pennie L. Anderson, Dan and Kim Short, Larry and Callie Epstein, Ann Berkley, Tom Palmer, Robin Tawney Nichols, Tim Speyer, Land Lindbergh, Hank Goetz, Ben and Penelope Pierce, Brooke Hunter, the Bennett Charitable Trust, Dave Odell, Chris and Autumn Carparelli, Phil Smith and Nancy Lee, Max and Kristina Davis, Sweetwater Ranch Properties, Bill and Melinda Berg, Bill and Linda Musser, Diane and John Decaro, Norma and Gary Buchanan, Nora and Chris Hohenlohe, Katie Van Dorn and Ray Johnson, the Margaret Kendrick Blodgett Foundation, Ben and Penelope Pierce, Roger Kiesling, Alex MacGrath, Mary and Baxter Brown, Roy and Susan O'Connor, Jim and Chris Scott, Mac and Pat Binger, Chris and Carol Hunter, Grant Alban, Michael Schechtman, Carol and Lenny Schweitzer, the Renaissance Charitable Foundation, and many others whose donations continue to arrive.

Elizabeth Richardson is a stalwart conservationist and we are especially grateful for her help and guidance.

Thanks go to four remarkable contributors to the text: first, staff members of the Confederated Salish and Kootenai Tribes provided a fine contribution exploring their conservation story and for that we thank (in alphabetical order) Tony Incashola Jr., Rich Janssen, Sadie Peone-Stops, and Thompson Smith, with assistance from Whisper Camel-Means, Les Evarts, Stephanie Gillin, Barry Hanson, Tom McDonald, and Shane Morigeau. Three additional strong contributions were written by Shane Doyle (Apsáalooke), Steve Running, and Todd Wilkinson.

The maps were expertly supervised by geographer Kevin McManigal at the University of Montana. Hannah Shafer applied her excellent cartographic skills throughout the project. Vital mapping work was also done by Alex Butler, Emma Montague, and Peter Scilla. Kevin League's photographs elegantly show the staggering beauty of Montana—he is an experienced conservationist with an artist's eye. Mark Sommer researched GIS data and kept things accurate. David Quammen provided early savvy and dot connections to the world of publishing. Thanks to all the people we interviewed, including Jon Roush, Joyce Chinn, Ellen Knight, Sarah Halvorson, Mary Hollow, Marie-Eve Marchand, Land Lindbergh, Hank Goetz, Roy and Susan O'Connor, Orville Daniels, Bill and Max Milton, Grant Parker, Rock Ringling, Sean and Kayla Gerrity, David Hartwell, Douglas Chadwick,

Hall & Hall, Sweetwater Ranch Properties, Fay Ranches, Mike Swan, Andie MacDowell, Mae Nan Ellingson, Spencer Beebe, Germaine White, and Ron Marcoux. Thanks to Hannibal "Andy" Anderson for his rancher's take on grizzlies. Gates Watson helped frame our message. Cathy Whitlock of Montana State University reviewed and improved the scientific accuracy of the "Nature's Montana" chapter. Paul Starrs read and enhanced the "Ranching" chapter. Spencer Beebe was generous in sharing his knowledge of fund-raising. Mark Milliorn helped make the "Crucial Role of Capital" chapter economically sound. Greg Tollefson's vast knowledge and experience were essential in writing the "Missoula Region" chapter. Many thanks, Greg. Wendy Ninteman, western director of the Land Trust Alliance, provided helpful land trust perspectives. Roy O'Connor, Hank Goetz, Land Lindbergh, and Jim Masar shared their deep knowledge of the Blackfoot watershed. Thanks to Montana Fish, Wildlife and Parks for its early addition of conservation easements to its conservation toolkit—a deep and lasting contribution. Tribal vice chair Tom McDonald (Salish) was kind enough to offer his wise and straightforward perspectives on conservation. Smoke and Thelma Elser shared their deep knowledge of wild places and horse packing. Thanks also go to Daphne Bugbee Jones, Erik Swanson, Gary Sullivan, Susie Lindbergh Miller, Elvin "Speed" Fitzhugh, Jill Cunningham, Pat Asay, Kim McMahon, Sally Moore, Nancy Johnson, Evan Denney, Tom Huff, James J. Parsons, Mary Lawler, and Libby Addington. We thank Dale Harris, an inductee to the Montana Outdoor Hall of Fame, for his work to protect the Alberton Gorge and the Great Burn landscape. We appreciate Jack Mauer, who gives more than he gets while outfitting in places like the Bitterroot Mountains. Dan Baily's fly shop in Livingston is a touchstone. Paul Roos knew every fish in the Blackfoot. Bruce Farling knows every fishery in the state. Senators Max Baucus, John Melcher, Conrad Burns, Jon Tester, and Steve Daines strongly supported conservation. Congressman Pat Williams was essential in creating the Rattlesnake National Recreation Area and Wilderness. Governor Tim Babcock walked the walk regarding land conservation.

Thanks to Monte Dolack and Mary Beth Percival for focusing their artistic gifts on land conservation through many project-based works. Clyde Aspevig's landscape paintings can illuminate the "Prairie" chapter. The art of Rudy Autio, Jaune Quick-to-See Smith, Larry Pirnie, Wendy Red Star, Kendall Jan Jubb, and so many others helps us see Montana in transformative ways. We humbly acknowledge that this story takes place in the homelands of Native American nations. Indian people are the original conservationists in Montana, with stewardship spanning more than 13,000 years. We are profoundly grateful for their wisdom, which resulted in the survival of a biologically intact landscape. The following are tribal names as we typically know them, followed by how the nations self-identify.

Arikara—Sahnish, Arikaree, Ree, Hundi
Assiniboine—Nakoda, Nakona
Blackfeet—Niitsitapi (Pikuni)
Chippewa—Ojibwe (Anishinaabe)
Crow—Apsáalooke
Gros Ventre—A'aninin
Hidatsa - Hiraacá
Kootenai—Ksanka
Little Shell Chippewa—Anishinaabe and Métis
Mandan—Numakaki
Northern Cheyenne—Tsetsêhesêstâhase (or Tsis tsis'tas) and So'taa'eo'o
Pend d'Oreille—Ql'ispé
Plains Cree—Ne-i-yah-wahk
Salish—Séliš-Ql'ispe
Sioux—Lakota, Dakota

PREFACE

In 2020, the three of us decided to try something bold—to tell the story of land conservation in Montana from Earth Day 1970 onward. This tale deserves to be told before the stories are forgotten. New projects are happening so fast that the vital lessons of history might be lost. The essential purpose of *Saving the Big Sky* is to inspire the reader to help conserve even more of Montana.

Robin Wall Kimmerer (Potawatomi) sees the work this way: "It is an intertwining of science, spirit, and story—old stories and new ones that can be medicine for our broken relationship with Earth; a pharmacopeia of healing stories that allow us to imagine a different relationship in which people and land are good medicine for each other."

This project was a wonderful challenge, one that required us to do interviews; gather data; remember successes, then failures; reflect, walk, map, write, and be humbled by the scale of what has been achieved. The book's structure came from the classic environmental history approach of using eras, followed by case studies to bring it all home. Dr. Shane Doyle (Apsáalooke) shows that Indigenous conservation began in the Ice Age and continues today. During this project there were countless Zoom meetings to maintain the flow. The Foundation for Montana History was kind enough to be our trusted home base.

Five years later, here we are. *Saving the Big Sky* is an uplifting record of land conservation spanning half a century. It is a clear expression of applied geography with great maps, striking photos, and (we hope) evocative writing. The book has a special focus: How are Montanans keeping biodiverse lands from being subdivided, developed, and otherwise transformed? Libraries are full of tomes about imperiled nature. *Saving the Big Sky* has a more optimistic purpose—to share how six million acres of beautiful, biodiverse private land were protected by good people. This remarkable story is needed in a time when positive news is hard to come by.

Does this make Montana the Last Best Place? Forgive us, but that sounds like the rest of the planet is beyond redemption. There are countless "best places" worthy of safeguarding. If not, what is the larger purpose of this book? Please take what is useful and apply it where you live. Invent new and better ways of taking care of landscapes.

Many dedicated people from all backgrounds are still out there conserving Montana—tribes, agencies, land trusts, national NGOs, corporations, and lots of property owners. This book is rich with people's names because they all deserve credit. Apologies to anyone we overlooked. Land conservation results from listening to each other, showing respect, and applying the right tools to get the job done. That sounds mechanical, but every project first existed in people's hearts and souls. In a way, *Saving the Big Sky* is a love story, one that reveals profound relationships between human beings and the land.

We are honored to share it with you.

Bruce A. Bugbee, Missoula, MT
Robert J. Kiesling, Helena, MT
John B. Wright, Missoula, MT

Crisp fall morning

CHAPTER 1

Land Stewardship

Geography is about searching to find our way. The landscape of Montana is a chronicle of human decisions etched into the good earth. Insights and illusions are revealed all around us—a telling map of our ways.

"A dreamscape," Debra Magpie Earling (Salish) calls it, with so many visions.

Reservations, ranches, subdivisions, mines, farms, cities, forests, and wildlife habitats. Each status and form of ownership reflects our aspirations, wisdom, and misperceptions. But what does "land stewardship" mean? To us, it means treating nature wisely so it will remain ecologically healthy and beautiful—so it will last. This is not just true in wild country; it includes ranches, river valleys, farms, and forests where we live and manage resources. Montanans love these cultural landscapes and call them home. Today, widespread land subdivision and housing development are the most powerful threats. "Land conservation" means protecting key lands from that outcome.

But how do we find the ethical balance between caring for the landscape and dealing with the practical realities of growth? How do we create a decent future while acknowledging private property rights and honoring the stories of those who came before us? Native American cultures counsel us to focus on the "7 Rs": Respect, Relevancy, Reciprocity, Responsibility, Rights, Reconciliation, and Relationships. These core values provide ample food for thought as we wrestle with change.

A Blackfeet proverb says, "What is life? It is the flash of a firefly in the night. It is the breath of a buffalo in the wintertime. It is the little shadow which runs across the grass and loses itself in the sunset." There is natural mystery in that, a humble experience of sacredness. Rosalyn LaPier, a Blackfeet professor, offers this: "Our landscape is saturated with environmental stories. Everything in Montana—every mountain, hill, river, creek, meadow, grassland, badland, animal, fish, tree, plant, bird, insect and even the night sky—has an Indigenous name and story."

Words reveal worlds.

The transformation of a Native landscape into the current State of Montana embodies all the complexities that define human beings. Violence and the quest for peace, kindness and cruelty, ruin and restoration, finding a way forward without whitewashing the past. We will not attempt to untangle all that. In this book we largely explore how private land is being conserved—protected from subdivision and housing development. That may seem like a narrow focus, but read on to see how critical that is for holding Montana together. Private lands are where individuals and families make personal decisions that determine the ecological and economic future of Big Sky Country. Public land conservation provides cores and connections, actions that reflect societal values and political processes through time. We offer fresh Montana case studies where useful. However, this book breaks new ground by stressing the overlooked role of private land conservation in maintaining ecological integrity and cultural bonds with place.

Land stewardship is the bedrock principle throughout. Stated plainly, land stewardship is being an enlightened caretaker of the natural world. Making certain that healthy landscapes survive for all who follow us, human and otherwise.
This noble relationship comes from both love and common sense since the terms "ecology" and "economics" are drawn from the same Greek root—*oikos*—"house." The place we live in. A hearth. In remote English dialects, "hearth" rhymes with "earth."

Montana generously offers a rich experience of life. Our sense of place impels us to *do something* to protect landscapes that sustain us physically and emotionally. But all events happen on land that is a Native homeland. This can feel like

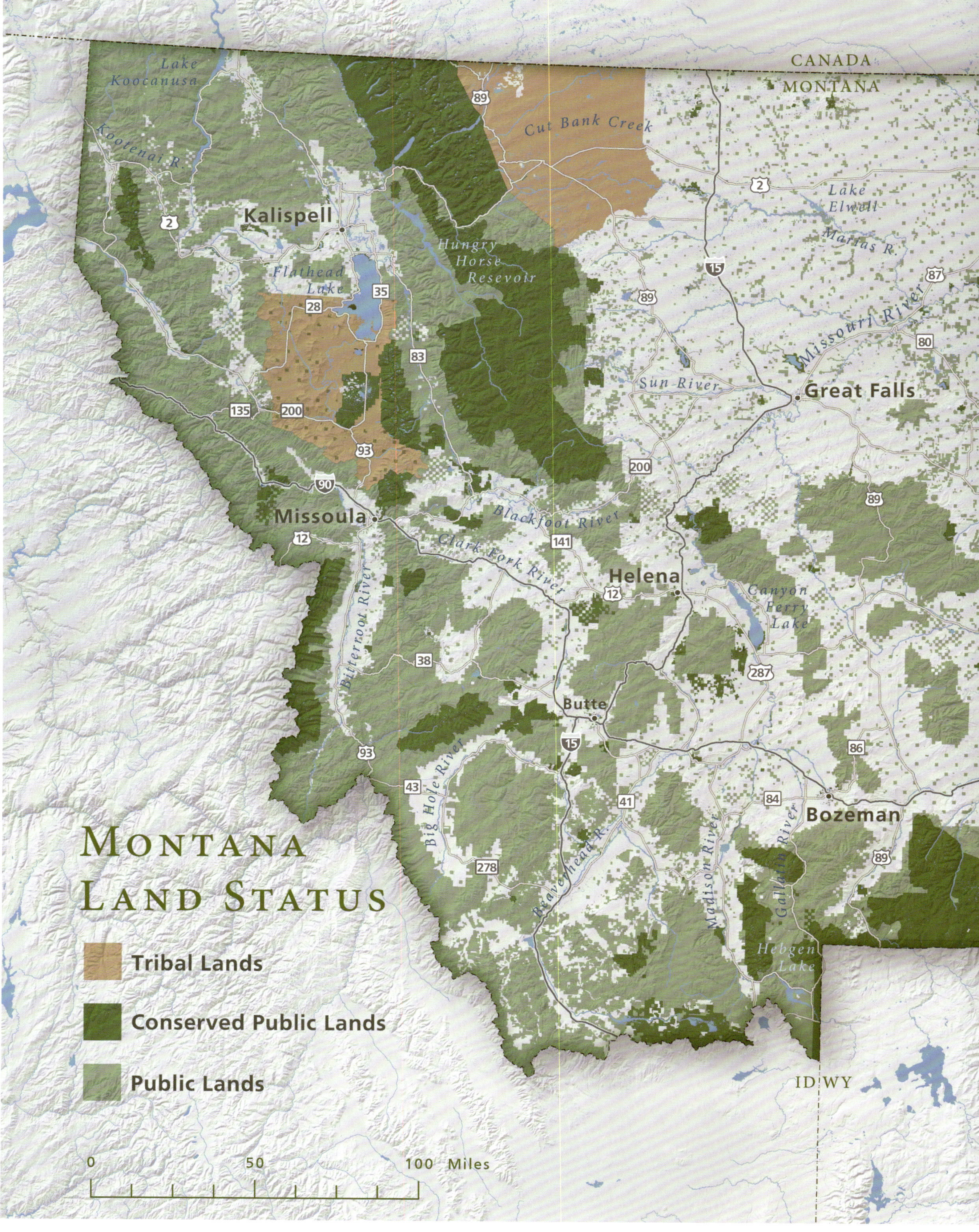
MONTANA
LAND STATUS
Tribal Lands
Conserved Public Lands
Public Lands
0
50
100 Miles
CANADA
MONTANA
Lake Koocanusa
Kootenai R.
Kalispell
Flathead Lake
Hungry Horse Reservoir
Cut Bank Creek
Lake Elwell
Marias R.
Missouri River
Sun River
Great Falls
Missoula
Blackfoot River
Clark Fork River
Helena
Canyon Ferry Lake
Bitterroot River
Butte
Big Hole River
Beaverhead R.
Madison River
Gallatin River
Bozeman
Hebgen Lake
ID WY

Montana: High, Wide, and Handsome.

Joseph Kinsey Howard

Prairies meet mountains on the Rocky Mountain Front

an ethical knot. In *Braiding Sweetgrass*, Robin Wall Kimmerer advises us to pursue two-eyed seeing—"learning to see with one eye with the strengths of indigenous knowledges and ways of knowing, and from the other eye with the strengths of Western knowledges and ways of knowing, and using these eyes together, for the benefit of all." Dr. Kimmerer suggests we use both Indigenous and Western perspectives to heal the land. But what do those diverse voices tell us about finding our way to stewardship?

Perceiving Montana as a series of overlapping historic eras can help. To begin, imagine a time before human beings entered this landscape. It's a cold sunny day and the Ice Age is waning. Withering glaciers are barely visible to the north. Prairies are born where tundra used to be, and vast herds of bison graze to the horizon with alert wolves standing close. There are elk beyond counting, mule deer and pronghorn, cutthroats lolling in creeks watched over by a sky lush with herons, hawks, and eagles. This is Nature's Montana.

Thousands of years ago, Native people arrived from Asia and spread across an unmarked continent. Following game, seeking warmth, expressing animist love, strong on their own two feet. They applied knowledge, adapted, learned, gave back, and found the right balance. In time they formed tribes we recognize who fully embrace existence.

Now flash forward to the nineteenth century in Fort Benton on the Missouri River, the terminus of navigation for steamboats. It is a time of commerce, ranching, and settling a "New World." But this was Native American country long before Lewis and Clark "explored" a fully known landscape and the paddle wheels churned.

Move west along the Teton River into the Golden Triangle, named during the Bonanza Wheat Boom of the early twentieth century. Small homesteads grew grain and prospered until the rains failed and many settlers left. Collapsing houses and barns are rough relics of tough times. This is now mostly ranch country, some of the best in the world. Part of the vastness is now called the Rocky Mountain Front Conservation Area, where peaks rise from the prairie and the sky stretches forever. Elk, grizzlies, swans, and a magnificence of other species thrive despite their troubles in other parts of America.

Montana is also part of the big world. Nuclear missile silos marked by cryptic roadside numbers are scattered across the plains. In the distance, land subdivisions ring the city of Great Falls, providing a decent place to live. But in every direction across Montana, rural housing developments expand into biodiverse ranching valleys. So does the use of voluntary tools like conservation easements to keep the land healthy. Much more on this coming up.

Head north past Heart Butte into the Blackfeet Reservation, the remnant of an immense homeland where Indigenous people defied repeating rifles and survey rods. In 1895, with the bison nearly gone, the Amskapi Piikani nation was conned into

OPPOSITE: A limber pine skeleton frames Castle Mountain in the Sun River drainage

ceding 800,000 acres to the United States. Their "Backbone of the World" became the eastern half of Glacier National Park. A sentinel stands to the north, Nínaiistáko, "Chief Mountain," a holy peak to Blackfeet people. All but its eastern flank is outside the reservation.

Enter the park and head west to Many Glacier Hotel. In 1915, this rustic chalet was built by the Great Northern Railway as a destination resort for visitors from around the world. Montana was advertised as "The American Alps" and tourism became an industry.

Hike along Swiftcurrent Lake to Grinnell Glacier, vastly smaller than it was a century ago. An aquamarine lake now holds the last meltwater. Both the ice and lake are named for a New Yorker named George Bird Grinnell who studied Nature here and promoted the establishment of Glacier as a national park. At the time, Montana newspapers firmly opposed it.

Now, follow a steep trail through blue and yellow wildflowers to Logan Pass, named for a Texan who was the park's first superintendent. His obituary read, "Indian Fighter and Noted Road Builder." But listen to happy voices in the parking lot, people awed by the sheer beauty of a wonderous circle of mountains. They are unaware of all that history.

After miles of heading south, the trail ends at a reddish cliff. These rocks are messengers from the foundation of time—nearly a billion years ago. Outcrops preserve ripple marks and mud cracks from the edge of a forgotten sea. Marmots and sky pilots show us the way until we reach a sudden crest.

Heavy Runner Mountain.

A sacred place, like so many, but this peak reminds us of the challenge of seeing Montana whole. This high ground is named for Blackfeet chief Heavy Runner, a man who was killed by US cavalry along with most of his encampment along the Marias River. That violence happened in 1870 when America had professional baseball and trolley cars.

If we're wise, we stop at the false summit of Heavy Runner. The last pitch is far too dangerous, off-limits for so many reasons. But from this height we have a humbling vantage point for seeing all the decisions about land and life that created Montana. Most of those choices remain.

If you are holding this book, you cherish Montana because it is blessed with Indian reservations, ranches, wildlife, rivers, farms, mountains, good communities, and profound beauty. There is no finer gift, but lots of new people are moving to Montana and building houses. Threats to the gift are rising. Native people know that story well.

But what does "Saving the Big Sky" really mean?

It does not mean opposing growth and economic expansion or criticizing agriculture and conservative points of view. Or caring about just public lands. Even though private lands are the focus of this book, "saving" does not mean promoting land-use planning regulations. This may seem confusing, but read on, study the maps, and gaze at the photos.

It turns out the most effective way to conserve Montana landscapes is voluntarily. By using financial incentives to keep private land open and in agricultural production. By promoting passive restoration as landowners keep the land in shape. As you'll see, free market capitalism is how millions of beautiful, ecologically healthy acres in Montana have been safeguarded from misplaced development. Land trusts, ranchers, tribes, NGOs, agencies, corporations, and landowners made this happen. This book explains how and suggests what more can be done.

Once again, the principle of *land stewardship* is our guide. Conservationist Aldo Leopold provides an elegant definition: "A thing is right when it tends to preserve the integrity, stability, and beauty of the biotic community. It is wrong when it tends otherwise."

Seems sensible.

Treating the land with reverence and humility is innate in many human beings, but we come to this wisdom from different cultural, political, and religious directions. Vine Deloria of the Standing Rock Sioux believes that Indian people "petition Nature for friendship . . . [it] is the next major step in understanding that our species must take." For Jews it is embodied in the ancient Hebrew phrase *tikkun olam*—repairing the world. For Christians *and* Jews, Genesis 1:1 says, "In the beginning, God created the heaven and the earth." The Bible says we have "dominion" over the land, but does that instruct us to subdue all of Nature? A deeper translation of dominion is "to lower oneself" and be humbled by our responsibilities.

Scripture seems to say, "Work *for* God's creation, not against it." Make a covenant with the divine. Give your word and live by it. Most ranchers and farmers deeply understand this; it is how they were raised, how they see their role in the world, but economic realities must also be understood and worked with. We are spiritual *and* material beings motivated by many things, including money and love.

Terry Tempest Williams puts it this way: "Loving the land. Honoring its mysteries. Acknowledging and embracing a sense of place— there is nothing more legitimate and there is nothing more true. This is why we are here."

The ancient poet Rumi agrees: "Out beyond the ideas of right-doing and wrong-doing is a field. I'll meet you there." In other words, *talk* to each other. Montanans might put it this way: "Respect me and my family, listen to what we have to say, and maybe we can strike a conservation deal." That is how so much ground has been saved under the Big Sky. By building relationships and being honest. What a remarkable story.

FURTHER READING

America's Public Lands: From Yellowstone to Smokey Bear. 2020. Randall K. Wilson. Rowman & Littlefield.

Braiding Sweetgrass: Indigenous Wisdom, Scientific Knowledge, and the Teaching of Plants. 2020. Robin Wall Kimmerer. Milkweed Editions.

Glacier National Park: The First Hundred Years. 2008. C. W. Guthrie. Farcountry Press.

Hearth: A Global Conversation on Community, Identity, and Place. 2018. Edited by Annick Smith and Susan O'Connor. Milkweed Editions.

Heroes of the Bob Marshall Wilderness. 2020. John Farley. Farcountry Press.

Homelands: A Geography of Culture and Place across America. 2001. Edited by Richard L. Nostrand and Lawrence Estaville. Johns Hopkins University Press (see chap. 14 about Montana).

Making America's Public Lands: The Contested History of Conservation on Federal Lands. 2023. Adam M. Sowards. Rowman & Littlefield.

Montana: An Uncommon Land. 1984. K. Ross Toole. University of Oklahoma Press.

Our Common Ground: A History of America's Public Lands. 2022. Adam D. Leshy. Yale University Press.

Preserving Yellowstone's Natural Conditions: Science and the Perception of Nature. 2022. James A. Pritchard. University of Nebraska Press.

Rightful Heritage: Franklin D. Roosevelt and the Land of America. 2016. Douglas Brinkley. HarperCollins.

A Sand County Almanac. 1986. Aldo Leopold. Ballantine Books (many editions exist).

That Wild Country: An Epic Journey through the Past, Present, and Future of America's Public Lands. 2019. Mark Kenyon. Little A.

This Contested Land: The Storied Past and Uncertain Future of America's National Monuments. 2022. McKenzie Long. University of Minnesota Press.

Wilderness and the American Mind. 2014. Roderick Nash. Yale University Press.

A Wilderness Original: The Life of Bob Marshall. 1986. Mountaineers Books.

Wilderness Warrior: Theodore Roosevelt and the Crusade for America. 2009. Douglas Brinkley. HarperCollins.

Yellowstone: 150 Years as America's Greatest National Park. 2022. Lew Freedman. Skyhorse.

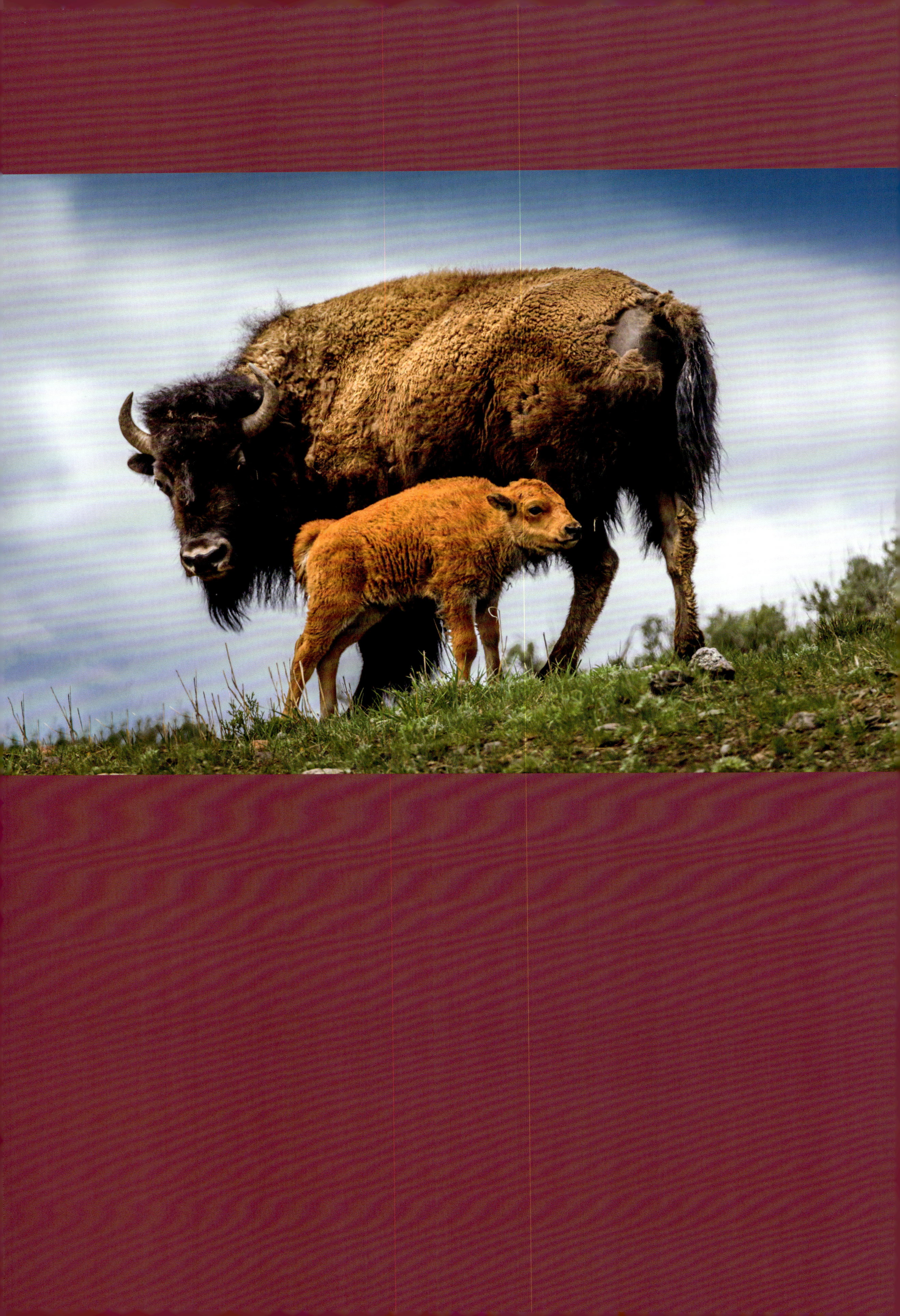

CHAPTER 2

Nature's Montana

A herd of bison graze beside Glacial Lake Musselshell in Eastern Montana. Flies buzz and gray clouds smother the sun as cold wind ruffles the bulls' thick brown manes. Young reddish calves nurse and nudge their patient mothers for milk and protection. One of them has twins. She stands taller and stronger against the wind—the genetic grandmother of all bison today. *Pte* is what Assiniboine people would call her one day.

A blue wall hems in meltwater from the glacial lake, a southern edge of the massive Laurentide Ice Sheet. Nearby, other megafauna loll and grunt, joust and ease away. Ancient horses, camels, elk, mastodons, woolly mammoths, ground sloths, musk oxen, and immense brown bears. The bison seem content, dining on grasses and spike moss. Safety, of a sort, in numbers. Wolves watch from the hummocks, intelligent eyes sizing up their chances. Saber-toothed cats prowl and snarl, proclaiming the upper hand.

It is the Ice Age in Montana; summer, but cold and wet enough to sustain glaciers. The Rocky Mountains were 70 million years old by then, and if you traveled west, you would recognize the ranges: Beartooth, Little Belt, Crazies; every ancestral height encased in ice. The Great Plains were also well formed, made from sediments washed down from the mountains, deposited in layers, and hardened into sandstone, shale, and limestone. This ramp of rock slopes east, high at the base of the Rockies, tilting toward a river that one day would be named the Mississippi. It is a landscape of coulees and buttes, wide expanses of tundra, and fierce rivers milky with rock flour from glacial erosion.

This is a glimpse of Montana 140,000 years ago, long before Native people arrived from Asia. In fact, human beings (*Homo sapiens*) had barely left Africa. Years later, our kind ventured across the world seeking food, shelter, adventure, and meaning—skin colors changing, but still the eternal us.

A hundred thousand years later another wave of bison crossed the Bering Land Bridge into North America. They entered Montana following an ice-free corridor between the Canadian Rockies and the continental glacier in Saskatchewan. Other creatures followed their tracks and diversified into species we know. Grizzlies, wolves, and bison spread across an ecologically promising world.

The "big three" species thrived in Montana. Grizzlies claimed top omnivore status, eating plants and animals, fresh or carrion, whatever was available and tasty. Wolves became the apex predator when dire wolves and sabertooths went extinct. Bison formed the basis of nearly everything—a keystone species and the apex herbivore, grazing and keeping ecosystems diverse and healthy, dispersing seeds, stirring the soil, cycling nutrients, and with their deaths, providing food to wolves and griz.

A coevolved ecological triangle. The enduring wisdom of a natural god.

The Ice Age began to wind down about 15,000 years ago as Montana warmed up. The reason why is debated. Cyclical shifts in Earth-sun relationships certainly played a role—changes in orbital shape and wobbles in the planet's axial tilt. As a result of these "Milankovitch Cycles," the 2.6-million-year-old Pleistocene Epoch consists of 17 alternating glaciations and deglaciations: cold and warm waltzing back and forth. There is no reason to think that dance has ended, so until further notice we still live in the Ice Age.

Geologists call the last 11,700 years the Holocene—"Wholly Recent"—perhaps out of fear (or now hope) that the cold will return. The Holocene has always been climatically dynamic, with rapid temperature oscillations. Today's climate change is nothing new in the geologic record, but its human cause and pace certainly are. As the world heats up, people hope for another natural

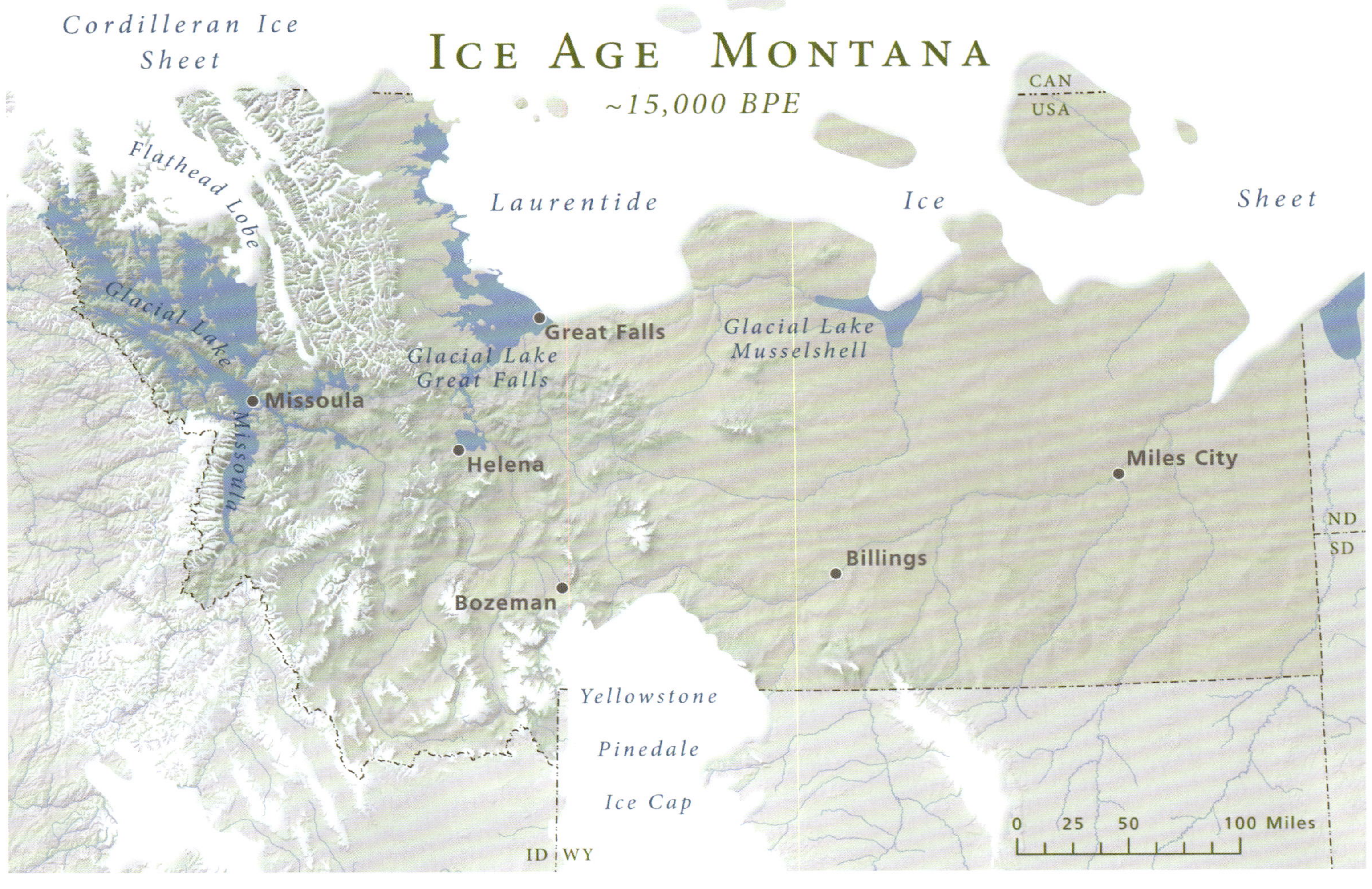

downturn. Problem is, the next natural cold cycle is thousands of years away.

Bison are the largest of the terrestrial fauna to survive the transition into the Holocene. Mammoths, mastodons, and other vegetarian monsters did not live through the shift caused by climate and habitat change. Bison survived and thrived mostly because of their adaptability. Native people created cultures that honored and depended on their "brother" on the plains. At that time, more than 30 million bison roamed North America.

A diverse diet helps explain why. Bison eat grasses, sedges, shrubs, forbs, and even small trees if they have to. They endure cold, heat, dry, and wet by migrating across large distances and leaving one habitat for another, seeking food, water, and cover on the hoof. Bison live in connectivity, teaching successful routes to their young. Ecologists call this suite of adaptive traits *plasticity*, which gives species like bison powerful advantages during environmental transformations. And in that is both a reason for hope and a cause for concern.

Now, imagine going back to the great transformation when the ice sheets and glacial lakes finally disappeared. The descendant of the ancient matriarch, the *pte*, stands on a butte. The warming prairie is generous with grass, spring-lush and fully formed. Prickly pear cactus is nearby, a new arrival her family has learned to avoid. Blue lupine and purple coneflowers scent the air. As the climate moderates, vast numbers of bison migrate north into Montana from the Southern Plains and spread.

The *pte* is in no hurry, but everything gets done—grazing, finding water, wallowing, sleeping. Her calf kicks up its heels and spins the way happy youngsters do. She stares north across the plains toward eroded breaks that lead down to a big river. This powerful country holds the bones of

OPPOSITE: "Twin Sisters" in northeast Montana's Makoshika State Park

Bison bull at ease

her ancestors. Piney ravines offer cover, but wolves lie in wait.

There is a new species in her world—human beings. At first not many, but now more and more. The *pte* escaped a stampede once, a panic caused by the humans that drove her cousins over a cliff. There were wails of pain, then silence, then butchered bodies. After that, the humans went away honored and fed by that sacrifice. She thinks of them as two-legged wolves. Her wariness grew, but with time the *pte* returned to life as it was before.

Such a grand life. Sweet grass to eat, a herd she grew up in, open air, and quiet. Bird music and bull bellows now and then. Hawk calls and prairie dog songs. This is the only world she knows or can imagine, a place she belongs to. The *pte* feels all this without trying. Her bond is ancient, innate, and unbreakable.

Love is biological after all.

Over her life she gives birth to many calves, teaching them what is necessary, smelling their bodies instead of giving them names. There are tough winters and mild ones, rainy springs and dry summers, but the prairie extends forever, a perfect union of bison and grass.

All this happens centuries before Indians acquire horses and White men introduce rifles. She lives a full natural life never knowing the horror of what is to come. Blessed by birth to live as bison always have, feeling a sense of place we can only dream of. North of here is a huge expanse of prairie on either side of a great river, later called the Missouri. Perhaps she would recognize it still.

Ecology and Empathy

The paleoecology of Montana is not just for science geeks. It offers an informed way of understanding the origin of current natural communities. It encourages empathy for each other and wild beings—our cousins with fur, feathers, and fins.

Paleoecology teaches profound lessons about land conservation—themes we will stress in this book. Large habitats are more diverse and resilient than small ones. Fragmented landscapes make it harder for species to live than intact ones. Water is vital. Migration corridors and connectivity give species a chance to adapt and survive. Corridors become filters when habitats contract and barriers when they disappear. But it's all relative. The "closing" of the Bering Land Bridge blocked a route for mammals but presented no problem for birds. The rising Bering Sea became a liberating corridor for marine life.

Pollen grains, the stuff that makes us sneeze, are a faithful recorder of the ecological history of Montana. Dr. Cathy Whitlock, an Earth scientist at Montana State University, is an expert in decoding fossilized pollen grains and charcoal in layers of lake deposits. She and other researchers extract sediment cores all over the state. The rings of living trees also inscribe changes in temperature, moisture, and fires in cellulose and carbon. These proxy measures of climate produce a robust record of how Nature's Montana came to be. It turns out that ecosystems we see today are relatively new and still evolving. They are more sensitive than we imagine and more vulnerable than we fear.

Montana's paleoecology is dazzlingly complex and full of surprises. Here's a sense.

As glaciers receded at the end of the last Ice Age, tundra vegetation was established, followed by spruce–fir–whitebark pine forests. Then came modern forests, wetlands, grasslands, and savannas in shifting patterns over space and time. Upper tree lines on mountains rose when it warmed and fell when it cooled—sometimes 3,000 feet of shift. In one study, Cathy Whitlock and her colleagues discovered something astonishing. From 850 to 1850, the Centennial Valley was cool and wet, reflecting the influence of the Little Ice Age. But similar studies show that Greenland is the warmest it's been in 1,000 years.

Natural science is a humbling way of learning.

By the early Holocene (11,700–6,000 years ago) we would recognize Montana's natural landscape, but its ecosystems were not analogs for the present. Things would look odd to us today, and some plants were missing.

Our beloved ponderosa pine became abundant in Montana only 10,000 years ago. Ponderosas arrived in Eastern Montana only 1,200 years ago. They spent the glacial period in the Southern Rockies before migrating north across tough ground. Douglas fir spread across the state from moist Western Montana refuges about 9,000 years ago. Just a few thousand years ago, junipers migrated in from the south. But lodgepole pines have dominated the Yellowstone Plateau since the ice melted, maintaining their rule by burning down every hundred years or so—a fire-dependent species in a landscape of infertile volcanic soils. In the heights, whitebark pines claimed subalpine habitats as early as 12,000 years ago, where grizzly bears ate their seeds and offered them back as fertile piles for dispersal. Today's warming climate leaves whitebarks vulnerable to disease, drought, and fire, reducing a major food source for grizzlies.

In 2017, an intense forest fire torched the subalpine forests of the Anaconda-Pintler Range. A beloved whitebark pine was killed in a once-in-forever burn. She fell face down into Upper Carpp Lake on a site that seemed immune to fire—a high-elevation peninsula, sandy soil, no fuel—but she died anyway. It was a stunning loss, staggering evidence that the natural world has changed, proof that botany can break your heart. Despite this event, the following facts remain ecophysiologically true out there in the forest.

Subalpine fire, Engelmann spruce, and whitebark pine are cold tolerant.

Lodgepole pine, juniper, limber pine, Douglas fir, and ponderosa pine are drought tolerant.

Western larch, alpine larch, grand fir, western hemlock, and mountain hemlock are moisture-loving Cascadian outliers found in Western Montana ranges like the Cabinets, Swans, Missions, and Bitterroots.

As the landscape gets drier and warmer, upper and lower tree lines are rising. Ecosystems are fraying at the edges, another reason to conserve land at all elevations. Each species of tree, mammal, reptile, shrub, flower, fish, and grass in Montana lives within its zone of tolerance for temperature and moisture, beyond which it struggles and cannot compete or live. Specialist species (endemics) like Missoula phlox have very narrow habitat requirements. Generalists like white-tailed deer make themselves at home all over the place. The narrower the need, the more vulnerable to extinction a species is.

Grizzlies were once a Great Plains animal forced into the mountains by settlement, but black bears remain at ease around us. Elk were a continental species until hunting and habitat loss pushed them into sanctuaries in Yellowstone National Park and the Sun River Wildlife Management Area. Wolves were everywhere until bounties were put on their heads, and they still struggle

for acceptance. By 1900, there were only 23 wild bison left, and the US Army was deployed to Yellowstone to protect them from poachers. They are now making a comeback on Indian reservations and nature preserves.

Despite settlement and growth, the biodiversity of Montana remains high. Charismatic megafauna do very well in the state: 140,000 elk, 300,000 mule deer, 220,000 whitetails, 15,000 black bears, and 5,000 moose. All of them are managed species. About 2,000 grizzlies live here, mostly in the Northern Continental Divide region, Cabinet-Yaak country, and the Greater Yellowstone. Wolves were reintroduced into Yellowstone in 1995, then packs invited themselves into the Ninemile Valley near Missoula and across the state. Returning 1,000 wolves to the Montana landscape didn't happen without controversy. Fear of attacks on people, predation on livestock, and impacts on deer and elk numbers combine to stir things up.

These wildlife successes occurred because of vast expanses of wild country, but mostly because people cared. First of all, Native people from a dozen tribes who treated the land well for millennia. Then people like Teddy Roosevelt, Margaret Murie, Bob Marshall, Lee Metcalf, Bearhead Swaney, Barbara Rusmore, Christine Torgrimson, Bob Munson, Earl Old Person, Cathy Whitlock, Tom McDonald, Wendy Ninteman, Curt Freeze, Bob Ream, Bob and Ellen Knight, Tracy Stone-Manning, Greg Tollefson, Jamie Jonkel, Gail Small, Smoke Elser, Robin Tawney Nichols, Mike Jimenez, Mary Hollow, and hundreds of other dedicated conservationists. As a result, all species seen by Lewis and Clark still live in Montana. These include big animals as well as endangered creatures like the black-footed ferret, wolverine, whooping crane, least tern, pallid sturgeon, bull trout, peregrine falcon, Canada lynx, yellow-billed cuckoo, western glacier stonefly, and northern long-eared bat. They all have a chance now.

But Montana used to have reindeer. Not in the distant past; in the twentieth century. Woodland caribou still occasionally drift south from Canada into forests near Yaak in Lincoln County, but they are now "functionally extinct" in the state. Herds once stretched across British Columbia, Montana, Idaho, and Washington. According to the Rocky Mountain Elk Foundation, "The decline is largely attributed to high mortality rates linked to habitat fragmentation, loss of old growth forest, and subsequent predation." Reindeer are now a federally protected species, something every child would praise, but providing habitat will take hard work. One of many challenges we face in "Saving the Big Sky."

The number of individuals in a species is important, but it doesn't exactly measure progress in land conservation. Habitats are also real estate owned and managed by individuals, government agencies, tribes, and nonprofit groups. Threats and opportunities arise every day. Stories emerge in upcoming chapters about how land conservation is faring in eight regions of Montana. Before that, to get our bearings about Nature's Montana, a basic geographic framework is needed.

Physiographic Regions

Geologists divide Montana into two landform realms: the Rocky Mountains and the Great Plains, with many divisions below that. Montana Fish, Wildlife and Parks uses large ecosystem names like Montane Forest, Intermountain Grasslands, Plains Grasslands, and Shrub Grasslands. Harder to envision. What distinguishes one grassland from another? Is it just location or is it species composition? Are we standing in a Montane Forest or a habitat type like Douglas Fir–Snowberry? The Institute on Ecosystems at Montana State University divides the state into four regions: Crown of the Continent (including Glacier), Greater Yellowstone, Upper Missouri, and High Plains. Dr. John Crowley of the University of Montana developed a "biophysical mapping" system for defining "geographic ecosystems"—areas with the same soil, slope, present vegetation, and US Forest Service habitat type. Effective, but far too detailed for use here. Individuals with deep experience in Montana develop their own mental maps of Nature based on experience, reading, and conversation. But it's an immense state, and one lifetime is insufficient to personally see it all.

This map illustrates our middle-ground approach. It shows five basic natural regions: large expanses of Montana based on landforms and vegetation.

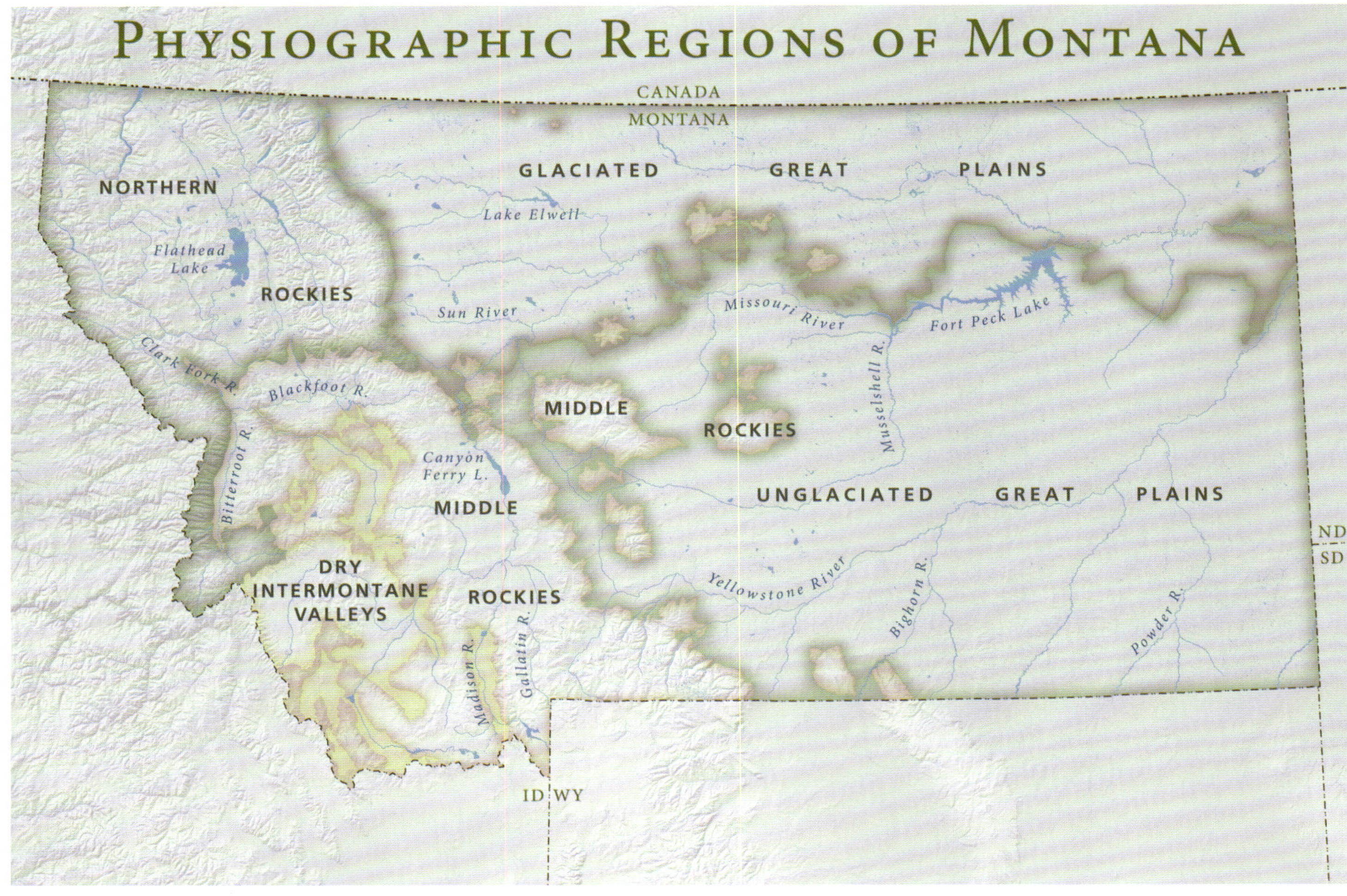

The Northern Rockies have narrow valleys between towering mountains that are home to diverse conifer species. These include trees usually seen in the Cascades such as western red cedar, hemlocks, and larches. The Mission Mountains, Swan Range, Cabinet Mountains, Rattlesnake Mountains, and Glacier are made of billion-year-old argillite (mudstone), quartzite, and limestone. The rocks are maroon, gold, and green. Glacial Lake Missoula filled and drained multiple times during the late Pleistocene when an ice dam blocking the Clark Fork River failed, then rebuilt. Dozens of ancient shorelines are visible on grassy hillsides across the region, and Flathead Lake is a remnant of that freshwater sea.

Yellowstone National Park is the centerpiece of the Middle Rockies, with the Beartooth Plateau, Gallatin Range, and Madison Range extending north of the park. Forest fires are common in dense lodgepole pine forests. The ranges decrease in density as you go east, where plate tectonics ran out of gas. The Crazies, Little Belts, Big Snowys, and Bears Paw Mountains are "sky islands"—conifer forests surrounded by plains.

The Montana Valley and Foothill Prairies cover treeless intermontane valleys between Rocky Mountain ranges. These semi-arid prairies are home to bluebunch wheatgrass, Idaho fescue, needle-and-thread grass, prairie Junegrass, blue grama, crested wheatgrass, western wheatgrass, and various species of sagebrush. Traditionally, this is ranch country. It's also where much of Montana's recent land subdivision is taking place, around cities like Missoula, Bozeman, Helena, Kalispell, and Hamilton.

The two Great Plains regions lie in the rain shadow of the Rocky Mountains. These dry rolling prairies are distinguished by slight differences in rainfall and the presence or absence of continental glaciation. The Missouri River forms a rough boundary between them. The Glaciated Great Plains lying to the north are a bit wetter than the Unglaciated. The Unglaciated Great Plains ecoregion has more badlands (like Makoshika State Park) and rugged hills. Five glacial lakes once spread south from the ice front: Cut Bank,

Great Falls, Musselshell, Circle, and Jordan. Their former beds are strikingly flat. Prairie pothole lakes and the Missouri River sustain many species of migratory and resident waterbirds in the Central Flyway. These include Canada geese, sandhill crane, American wigeon, northern pintail, green-winged teal, bufflehead, wood duck, and tundra swan. Most landscapes are a mosaic of ranches and wheat farms. Unplowed prairies are a swirl of western wheatgrass, blue grama, prairie Junegrass, needle-and-thread, and colorful wildflowers.

The Montana Pole

The North Pole and South Pole are familiar, the Third Pole is not, and the Montana Pole offers a fresh perspective. The Third Pole encompasses the glacier-covered Himalayas and Tibetan Plateau—the source for the Yellow, Yangtze, Mekong, Indus, Brahmaputra, and Ganges Rivers. Billions of people depend on this water for irrigation and hydropower.

The Montana Pole extends from Glacier to Yellowstone, embracing the high Rocky Mountains in both physiographic regions. These ranges capture moisture from Pacific storm fronts, store it over the winter as snow, and then release it in the spring to dozens of rivers on both sides of the Continental Divide. Ag operators, fish, bugs, birds, and anglers are delighted, but dependent. The Montana Pole is a vital source of water for a swath of the continent from Washington to Missouri, but its future is uncertain.

In 1850, Glacier National Park had 80 named ice sheets like the Grinnell, Sperry, and Jackson. Kootenai people called the area "the place where there is a lot of ice." Today, the park only has 26 glaciers left, with the remaining 54 either inactive (not moving) or gone. One day this may be called Glaciated National Park.

Climate change in Montana now creates wild swings in precipitation from year to year and often elevates summer temperatures above 100 degrees Fahrenheit. Watersheds in the Montana Pole feel the strain. Every spring we check snowpack statistics in hope of good news. We used to pray for summer to arrive, but now it can be a season to endure. Not every year, certainly, but too often July and August are a scorchy test of our patience. In recent decades, forest fires have grown bigger and more intense, start earlier, and last longer, filling the air with choking smoke for weeks at a time. Fire is required in Montana ecosystems for nutrient cycling and plant succession, but tree-ring data show that ground fires used to be the norm, instead of the immense stand-destroying blazes we often see today. Yellowstone's lodgepole pine forests are the exception—a fire-dependent ecosystem evolved to burn dramatically.

Fires, snowpacks, and river flows measure climate better than any theory or argument. Nature is the truth. However, stream gauge data reveal the importance of the Montana Pole for irrigation and aquatic and riparian habitat. The news is unsettling.

Nobel Prize–winning climate scientist Dr. Steven Running substantiates three trends: (1) Montana is getting drier—"aridifying," he calls it; (2) Montana is warmer than it was, with baking temperatures all too common; and (3) wildfires are now more intense. When Lewis and Clark crossed back into Montana in June 1806, they were blocked by massive snowfields. Today, June on Lolo Pass is a time to photograph blue camas.

The Montana Pole is vital, but vulnerable.

The Missouri River is born there and in a good year provides one-fifth of the flow of the Mississippi. The undammed Yellowstone typically contributes a big share of the Missouri's discharge. The Snake technically starts in the Wyoming core of the park, but perhaps we can be forgiven for including it here. The Salmon River sends water from the Bitterroot Mountains to join the Snake. The Clark Fork seems modest, but it has the same average discharge as the Colorado River, which tenuously supports millions of desert-loving people. The Flathead begins in Glacier Park, joins the Clark Fork southwest of Flathead Lake, and together they carry meltwater into the Columbia and out to sea. Then there's Triple Divide Peak—part of its snowmelt flows into the St. Mary River, then on to Hudson Bay and the Arctic Ocean. The rest is split between the Atlantic and Pacific Oceans. Montana is consecrated with diverse pure water, and even its tributary rivers would be big news in most states: the Blackfoot, Bitterroot, Big

Hole, Jefferson, Gallatin, Madison, and Milk. The Smith, Stillwater, St. Regis, and Swan. Agriculture uses its share, and the remainder sustains everything else. Including the state's transcendent trout fishery.

Brook trout are covered with pink and yellow speckles. Small and frisky. Brown trout look like sharks in comparison. Eighteen inches long, painted with black, brown, red, and orange spots surrounded by halos. Rainbows shimmer with iridescent, elegant beauty. Easy-to-fool cutthroats have red slashes on both sides of their mouth. Golden trout are ever so rare. Bull trout are so precious they must be released wherever they're caught. The species was once so common the Salish people called Missoula the "Place of the Small Bull Trout." Regardless of type, one of the grandest experiences in Montana is watching a trout rise along the surface of a river. Rainbows and browns are European exotics, and brookies were introduced to Montana in 1889. Whatever the species, we visit Montana's trout waters to fish, float, and seek peace. To watch osprey, herons, and eagles pull trout from the water as they have done for millennia. To skip a rock or soak our tired bones, always knowing that water brings life to all.

Twelve waterways have been partially or fully covered by Montana's in-stream flow law. In 1969, the state reserved unappropriated water rights for fish and wildlife habitat in Blue Ribbon fisheries like the Blackfoot, Missouri, Gallatin, Madison, Smith, and Flathead Rivers. Rock Creek is protected from dewatering from its headwaters in the Pintlers to where it merges with the Clark Fork. Big Spring Creek in Fergus County is safeguarded from its mouth up to the state fish hatchery. Dave Odell, working for Trout Unlimited, purchased water in Painted Rocks Reservoir that is released every August, when the Bitterroot River used to go dry.

But the 2017 Montana Climate Assessment showed that mountain snowpacks have declined significantly since 1980. Droughts are natural in Montana, but rising temperatures will deepen dry times. When stream flows decline, everything is affected.

The Montana Pole is the fluvial heart of the state. If it weakens, the land dries up.

Caring and Conserving

Montana fills our eyes with beauty and our souls with hope. Beyond maps, species, and biogeography, what drives people to conserve is a love of the land—a truth worth repeating. As you'll see, practical tools are needed to conserve landscapes, but love and fidelity to the Earth make it all happen. Biologist E. O. Wilson calls it *biophilia*. This chapter is a science-based sketch of Nature's Montana, nothing more. Maybe these words will evoke personal memories that you can add to the tale.

We experience Montana as breathtaking vistas and quiet places, as a home with deep cultural and soul connections. Montana ranch families are deeply allied with the land after five generations. Try to imagine how bonded Native people are after 600 generations. They are *indigenous* to this place, and belonging is everywhere: riding on a trail in the Bob Marshall Wilderness—"big country" in the words of Germaine White (Salish); seeing ruts from travois up Alice Creek in the Blackfoot; attending a powwow on the same ground your ancestors did; seeing land in seven Indian reservations and the embracing beauty around them. Yellowstone Park was shared by many nations, a place to hunt and walk in beauty. Meadows in the Sweet Grass Hills are sacred to many tribes. Crow people call the Crazies Awaxaawippíia, which translates as "ominous mountains." It was here that Chief Plenty Coups had a vision that cattle would replace bison on the plains. The Crow still come here to fast and pray.

Non-Indians can never become Indigenous, but in the words of Robin Wall Kimmerer, "With enough bonding to place, you can become naturalized." Maybe we love a reach of a river where we caught and released the same trout or a high meadow where we shot our first elk. That blue heron nest is used by the same birds every year. In grief and joy, we lean on a tree for solace, a Medicine Tree. Maybe our sacred place is where we scattered the ashes of a loved one in a side valley they adored. My marriage proposal happened on that bluff, or my child spoke their

OPPOSITE: Metate meets the sky atop Badger Pass in southwest Montana

first word beside that spring creek. The flexing curve of a dry fly line on moving water. The sight of scuffed boots on granite; the rise of dust on a backcountry trail. Feeding desperate cattle during a blizzard, helping a neighbor do the same. Memories of a youthful climb up a rugged peak. Mycorrhizal fungi beneath our feet communicating through soil from tree to tree—sharing water, minerals, and food, warning against harm, inventing cellular knowledge. The rare sight of a mountain lion, the daily sight of deer. Up ahead is a pinegrass glen where we made love and meant it. There is my favorite camping spot. That cobble bar is where we played with our beloved rescue dog. Stories about families and friends resonate everywhere. Montana is a library of evermore love—a topography of spirit.

Places remind us to be kind. They encourage us to remember history and its tough lessons. By conserving land, we show reciprocity and honor a relationship worthy of the name.

Montana is our shared home. Nature is our genealogy.

FURTHER READING

American Bison: A Natural History. 2010. Dale F. Lott. University of California Press.

Biogeography: An Ecological and Evolutionary Approach. 2010. C. Barry Cox and Peter D. Moore. Blackwell Scientific Publications.

Biophilia: The Human Bond with Other Species. 1984. E. O. Wilson. Harvard University Press.

Ecoregions of Montana. USFS, USDA, EPA, and others. https://gaftp.epa.gov/EPADataCommons/ORD/Ecoregions/mt/mt_front_1.pdf.

Ecoregions: The Ecosystem Geography of the Oceans and Continents. 2014. Robert. G. Bailey. Springer.

Elk of North America: Ecology and Management. 1982. Jack Ward Thomas and Dale E. Toweill. Wildlife Management Institute.

Fire Scars. 2023. John B. Wright. University of Nevada Press (a novel).

Flora of the Pacific Northwest: An Illustrated Manual. 1998. C. Leo Hitchcock and Arthur Cronquist. University of Washington Press.

Forest Habitat Types of Montana. 1977. Robert D. Pfister, Stephen Arno, and others. USDA General Technical Report INT-GTR-34.

Mountain and Shrubland Habitat Types of Western Montana. 1974. W. F. Mueggler and W. P. Handl. USDA/USFS.

"Past Warm Periods Provide Vital Benchmarks for Understanding the Future of the Greater Yellowstone Ecosystem." 2019. Cathy Whitlock and Steve Hostetler. *Yellowstone Science* 27: 72–76.

Ponderosa: People, Fire, and the West's Most Iconic Tree. 2015. Carl E. Fielder and Stephen F. Arno. Mountain Press.

Roadside Geology of Montana. 2020. Donald W. Hyndman and Robert C. Thomas. Mountain Press.

OPPOSITE: Birdtail Butte in west-central Montana bathed in Milky Way light

CHAPTER 3

Native Voices

By Dr. Shane Doyle, Apsáalooke

Tribal Nation Building in the Twenty-First Century

Native American leadership and participatory collaboration in land and wildlife conservation efforts across the State of Montana have gradually become more common over the past three decades, and that trend is at an all-time high. Tribal governments and Tribal nonprofits, as well as individual Tribal members from across the state and throughout Indian Country, have begun to engage in partnerships with conservation nonprofit organizations to work toward common goals of preserving the natural ecosystem and protecting natural resources in a way that honors the balance between people and the environment. There are many examples of this Tribal-NGO collaborative conservation model, and this book will consider some that have concluded successfully and others that are being built and fortified through ongoing communications and a history of trustful interactions. Every Tribal community in Montana has representatives working alongside groups like The Nature Conservancy, The Wilderness Society, Wild Montana, the American Prairie Reserve, the Greater Yellowstone Coalition, the National Parks Conservation Association, and others at the local, state, and federal levels to utilize Native wisdom and strength to conserve what's left of Montana's wildlife and wild landscape. This culture of collaboration with conservation nonprofit groups was essentially nonexistent between 1970 and 1993, but a series of notable congressional acts began to turn the tide by expanding Tribal power and authority over Tribal and public lands and cultural resources.

Federal Indian policy has always wielded a huge economic and cultural impact on Native people in Montana and across the nation, and it has shifted dramatically and repeatedly since the first era of treaties began in the late 1700s. The current policy of Tribal self-determination began in 1970 and marked a pendulum swing away from over 20 years of a policy based on efforts to permanently terminate Tribes. Federal agencies used a variety of social assimilation strategies to implement the policy of termination, including Tribal relocation, where individuals were removed from reservations and bused to urban areas so that they could assimilate into the larger "melting pot" cities. This policy era, which started after World War II, served to economically destabilize and socially disintegrate Tribal communities throughout the United States and led to continued high rates of poverty on Montana's seven reservations. The self-determination era that began in 1970 slowly began to change the laws and policies that were so effective at devastating Tribal cultures. Some of the legislative measures instituted during the self-determination era include AIRFA, the American Indian Religious Freedom Reformation Act (1994); and NAGPRA—the Native American Graves Protection and Repatriation Act (1990). More recently, the Biden administration issued an executive order for all federal agencies to consider traditional ecological knowledge when developing strategic plans for natural resources and public lands management. As with all federal Indian policy eras, it takes a generation before the true effects of the policies begin to become visible and evident both within and outside Tribal communities.

Twenty-first-century Montana Tribal communities maintain a deep sense of ceremonial reciprocity toward their homelands, despite a host of contemporary social challenges. Over hundreds of generations, the traditional Indigenous way of life left little trace of resource waste or permanent infrastructural impediment to wildlife migration. Wildfires were common in the mountains and

OPPOSITE: South Fork Teton River carries spring runoff

KTUNAXA ?AMAK?IS
BLACKFEET
RESERVATION
ROCKY
BOY'S
RES.
KALISPEL
SCHITSU'UMSH(COEUR D'ALENE)
FLATHEAD
RES.
SALISH-
KOOTENAI
NIITSÍTPIIS-S
SALISH-FLATHEAD
NIMIIPUU
(NEZ PERCE)
LEMHI-SHOSHONE
SHOSHONE-
BANNOCK
EASTERN
SHOSHONE

MICHIF PIYII (MÉTIS)

FT. PECK RESERVATION

FT. BELKNAP RESERVATION

ASSINIBOINE

OČHÉTHI ŠAKÓWIŊ

HKOII (BLACKFOOT / NIITSÍTAPI)

MANDAN, HIDATSA, AND ARIKARA

APSÁALOOKE (CROW)

CHEYENNE

ITAZIPCO

SIHASAPA

CROW RESERVATION

NORTHERN CHEYENNE RES.

Tribal Montana

Ancestral Territories and Current Reservation Boundaries

on the prairie, so the landscape burned with low intensity and kept the ecosystem balanced and vibrant. Living in close harmony with the natural elements, Native people in Montana likely never considered the possibility that they could hunt the bison to extinction, take all the fish out of the rivers, or pick all the wild plants and berries from the land. Instead, their restrained and nonwasteful approach to hunting and gathering reflected the environmental values embedded within their languages and ceremonial ways of life. Tribal traditions that cherish and celebrate the purity and sacredness of water are ubiquitous, with the sweat lodge ceremony being the most common. Ceremonies and everyday practices such as not putting human or other waste into rivers, lakes, and creeks were ecological credos that bonded the people together in ways that few other things could. Nearly all those ancient ceremonies and traditions that recognized the sanctity of the natural world are still utilized today, so the foundational understandings of Tribal identity remain alive, albeit smaller than they once were.

An Ancient Sense of Conservation

Long before Montana became a state and its modern borders were drawn, Native people recognized the place as a central and pivotal zone unto itself, where the rivers of the Earth started and flowed in opposite directions. The state's broad and diverse landscapes contain vibrant ecosystems that made it a bastion for the Plains Indian way of life. Over the past 13,000 years, the archaeological record shows an ebb and flow of people, but the last 2,000 years brought dozens of diverse Tribal nations here, and many stayed and continue to maintain their presence on their homelands.

Before colonial forces began to dramatically shift the populations of people and wildlife in the nineteenth century, Montana's Indigenous communities represented a hinterland for some of the largest language families in North America. To the north, the Blackfeet and Cree speak an Algonquian dialect that comes from the Great Lakes. In the east, the Apsáalooke (Crow) and Lakota speak Siouan languages from the Mississippi Valley. Down south, the Shoshone speak a form of Uto-Aztecan, which is centered in Mesoamerica. West of the divide, the Bitterroot Salish trace their dialect to the Columbia River and Pacific coast. Remarkably, the Kootenai language is the only tongue isolated to Montana, although Native people understand their relationships to their homelands as being one of continuity and reciprocity since time immemorial.

Montana's ancient character as a blessed, but at times hostile land has always kept civilization at bay, and that holds true into the twenty-first century. The winter of 2022–2023 set records for cold and snow, and even wild animals perished in large numbers. Mesoamerican farming and agricultural traditions that swept across North America never took root in Montana because Native people recognized that Montana is not a place that can be tamed or cajoled; rather, it must be embraced and accepted. Ancestors of today's Apsáalooke nation successfully grew corn, beans, and squash for a short time along the far eastern portion of the Yellowstone River near present-day Glendive. However, the site was abandoned about 500 years ago as the Apsáalooke people moved farther west toward the headwaters of the Yellowstone River and became full-time hunter-gatherer-traders. Their collective experience and the wisdom they took from the land is reflected in an old Tribal place-name near present-day Springdale—Where the Corn Died.

Along with the extreme climate, Montana's topography is also distinct from that of other zones on the Great Plains because its Continental Divide area is diffuse and accessible, making it a thoroughfare for Native peoples both east and west of the divide who ventured across cultural zones to hunt, gather, and trade. The snowcapped Rocky Mountains connect directly eastward to a sweeping prairie interspersed with clear, cold rivers and bursting with island mountain ranges. These magical and mysterious mountains dot the open plains and harbor essential lodgepole pine trees as well as diverse edible and medicinal plants, fresh water, chert rock quarries, winter campgrounds, ceremonial areas, and rendezvous sites. All these blessed resources are not available on the eastern plains of Colorado, Kansas, or Nebraska. North and

South Dakota, Canada, and Wyoming all fall short of the dynamic place we know as Montana. The Apsáalooke chief Sore Belly expressed a similar sentiment in the 1830s: "The Crow Country is in exactly the right place. Any direction you go from here is worse. In the north the winters are long and cold. In the south it is too hot and not much water. To the west there are no buffalo and all the people eat fish and pick bones from their teeth. Far to the east the water is warm and muddy, and our dogs won't drink it. Here in the Crow Country, we have the best of everything. High mountains with snowy peaks, cold clear water, good grasslands with many buffalo, elk and deer, and lots of good wintering sites. The Creator put the Crow Country in the exact right spot."

A Deep History of Diversity and Unity

Twenty-seven tribes recognize Montana as part of their historical homelands, and they all share one special ancient relative whose short life there stretches back over 12,600 years. Along the Shields River, near the town of Wilsall, the oldest known burial in the Western Hemisphere was discovered in 1968. Genetic sequencing completed on the remains in 2013 revealed that the two-year-old boy who was laid to rest there is a common ancestor to nearly every Indigenous person in both North and South America. This beloved child was a member of one of America's First Families, and they honored him with a profound collection of priceless, exquisitely crafted arrowpoints, knives, and scrapers. The love and dedication shown to this Anzick Clovis child through his burial reveal the ancient values of Indigenous people of the Northern Plains. Though the child's family lived through Ice Age climatic conditions, faced down apex predators such as short-faced bears and saber-toothed cats, and harvested 11,000-pound woolly mammoths, they still mustered an astonishing collection of their best possessions to lay a toddler to rest. The historical significance of the First Family's discovery cannot be understated, and they remain the only family on the continent to ever be directly associated with Clovis artifacts.

The values exhibited in the Anzick burial reflect a community determined to honor the sanctity of the circle of life. These traditional ways of knowing and being are evident throughout history and in contemporary Native American communities. The Lakota Sioux phrase *mitakioyasin*, translated as "thanks and blessings for all my relatives," is a ceremonial saying that lies at the heart of Plains Indian philosophy and worldview and has guided Native communities for as long as the Tribes can remember. As dynamic and mobile communities, Plains Indians lived a ceremonial way of life and measured their wealth not through material possessions but through their rich connections to the world.

I am a Native person with deep familial roots in the region and was on a research team from the University of Copenhagen who discovered that the Clovis boy was one of my ancient relatives. In him, I see the unbroken circle of love and wisdom that connects Native people today with our ancestors and the landscapes we recognize as home. I helped coordinate the reburial and repatriation of the ancient boy, back near the exact site where he was disturbed in 1968. This ceremonial reburial was attended by Tribal representatives from throughout the Yellowstone region, and the gathering would have looked familiar to the First Family; close relatives and community loved ones coming together to pay tribute, reflect, and provide solace to one another. The Native people who attended the ceremony that day were completing a full circle back to the sacred and magical Yellowstone region.

From the very beginning, the Tribes of Montana lived a sustainable way of life based on hunting, gathering, and trading with each other throughout a seasonal cycle of movement. After spending months in wintering campsites, Tribal communities would move from those locations along the rivers during the springtime when flooding, mosquitoes, and grizzly bears gave people good reasons to move to higher ground and to nearby rendezvous sites to harvest plants and trade with their neighbors. This ancient lifestyle of Tribal communities walking seasonal rounds existed for millennia before horses and created a culture of consistent and beneficial interaction that allowed Native people to share ceremonies, music and dance traditions, hunting and gathering strategies,

Chewing Black Bones, Blackfeet Reservation Campground

ancient star stories, and myriad other cultural innovations. This age-old cosmopolitan interaction eventually spawned one of the world's great languages, Plains Indian Sign Language. The world's only noncolonial lingua franca, Plains Indian Sign Language connected a communication network with extensive knowledge. As with many other fragile oral traditions, the common use of this sign language diminished greatly with the onset of the twentieth century and modern life.

Treaty Era

This period began benignly in Montana in 1825, with the signing of the Friendship Treaty along the Yellowstone River, which included the Mountain Crow Band of the Apsáalooke Nation. By this time, smallpox epidemics had shattered Tribal populations throughout Montana, with fatality estimates as high as 75 percent. As diseases shredded Tribal balances of power, Tribal militarism came to the fore and was the only means to negotiate with the US Army. The 1825 treaty had little relevance at the time because it assigned no reservation boundaries and required no transactional agreements between the Tribes and the federal government. Those sign-and-trade agreements came later, with the first treaty of consequence in 1851. That first Fort Laramie Treaty divided the entire state into Tribal reservations and common hunting grounds and became known by the Crow people as "The Time of Living Inside the Imaginary Lines." The negotiations were precipitated by the 1849 California gold rush and encouraged by the legendary mountain man Jim Bridger, who understood that the Plains leaders would much rather negotiate for peace than fight a costly war. That gold rush was the first major invasion into Plains Indian Tribal homelands, and it caused environmental destruction that rippled across the region. Massive wagon trains with tens of thousands of oxen and horses despoiled essential winter campsites and disrupted vital hunting grounds along the riverways on the prairies east of the Continental Divide, pushing Tribal warriors to fight for their communities' literal survival. Luckily for the Montana Tribes at the time, the South Pass went through Wyoming, thus sparing the Big Sky state the inevitable onslaught of mining culture until 1864, when the Virginia City gold rush established the Bozeman Trail.

Although the 1868 Fort Laramie Treaty is remembered as a victory for Red Cloud, all the Tribes in Montana suffered land loss from the subsequent agreement, which may have closed the military forts but never came close to slowing the wave of immigrants to the Montana goldfields. Within 20 years of the discovery of gold in Alder Gulch, Montana Tribes suffered through the worst starvation winter on record. During the winter of 1884, an estimated 2,000 Native people in Montana starved to death, mostly on the Blackfeet and Fort Peck Indian Reservations. This infamous starvation winter is well remembered in the Blackfeet community and happened just prior to the dawn of the boarding school era, which coincided with the highest rates of child mortality ever recorded in Native communities. Disease, warfare, starvation, and boarding schools were the dominant forces impacting Tribal communities as they entered the twentieth century with only a fraction of their original population and land base. Federal Indian policies aimed at dismantling, disintegrating, and destroying Native culture and communities continued with only a few caveats until the era of Native American self-determination began in 1970. Unremarkably, the social and familial damage done to Native people over a century of colonization has brought severe consequences to contemporary communities. It will likely take decades of decolonization before most Tribal nations return to their former independence.

Modern Culture

Until the late twentieth century, the concept of conserving land was not a top concern for most Tribes, as other modern issues took prominence in Tribal communities. As Montana Tribal communities survived the many ebbs and flows of federal Indian policy throughout the past 150 years, some Tribes, like the Confederated Salish and Kootenai Tribes (CSKT), achieved significant steps toward modern nation building and began to impose their cultural environmental ethics within their own reservation, as well as outside reservation boundaries. The CSKT are leading the state in land, water, and wildlife collaborative conservation efforts, and other Tribes in Montana continue to address ongoing environmental problems in their rural communities just as they seek to protect historical homelands both nearby and far away from their reservations. Just as the CSKT achieved Tribe State EPA status in the 1980s when government regulations allowed them the opportunity, the Northern Cheyenne achieved the same status for their air quality regulations, forcing off-reservation entities to adhere to higher environmental standards for both air and water. Colstrip's soaring coal smokestacks are a towering testament to the impact the Northern Cheyenne nation had on air quality for their entire region.

The twenty-first century has seen a rise of Tribal influence toward conservation of lands outside reservation borders, which is reflected in the Crow Nation's involvement in and impact on the formation of the Custer Gallatin National Forest's 30-year management plan in 2019–2020. Through face-to-face meetings, letters of request, and a media campaign that included a short film about the sacredness of the Crazy Mountains in the Apsáalooke culture, the Tribe succeeded in influencing the Forest Service plan to designate over 30,000 acres of wilderness and backcountry lands within the Crazy Mountains. The Crow and other Montana nations are currently negotiating with federal agency officials, including leaders of Yellowstone and Grand Teton National Parks, to establish comanagement agreements that empower Tribal communities to impress their wisdom and traditional values on federal land management. The Blackfeet Nation celebrated the cancellation of the final oil and gas drilling lease issued in the Badger–Two Medicine area. The preservation of the Badger–Two Medicine was a nearly four-decade-long struggle for justice that buoyed the Blackfeet and other Indigenous nations throughout the state to never give up in their efforts to preserve their sacred landscapes.

Reclaiming Traditional Homelands

The 150th anniversary of Yellowstone National Park in 2022 provided an opportunity for the park to commemorate the Native history and continued presence in the area, which has long been understated and misunderstood. I helped lead the installation of the first-ever All Nations Teepee Village during that summer, and by all accounts it was a great success. Working collaboratively with the Bozeman-based nonprofit Mountain Time Arts and Tribal representatives from every reservation in Montana and Wyoming, the Teepee Village brought contemporary Native voices and a variety of art, Tribal culture, history, and tradition from 13 Tribes. Dr. Ren Freeman (Eastern Shoshone) helped lead the installation and coined the term "Yellowstone Revealed" to underscore that although Native people may have disappeared from the story of Yellowstone, we have never really left the place and continue to go there every year for ceremonial and other purposes. The Teepee Village was over a year in the planning and has the potential to grow to as many as 27 teepees—one for each Tribe who calls the park their homeland.

The new success of the Teepee Village concept in Yellowstone National Park comes just as the park seeks to change and upgrade its bison policy and is working collaboratively with Tribal representatives, other government agencies, and nonprofit groups, both Tribal and non-Tribal, to find a solution to the ongoing problem of bison slaughter at the park boundary. The common threads that link bison to the park, to federal agencies, and to nonprofit organizations are Tribal governments and Tribal representatives. Tribes and Tribal representatives have begun to recognize the power of their position when they establish a unified vision and working strategy with other Tribes, federal agencies, and NGOs. This theme has been strengthening and recurring since the early 1990s when an inter-Tribal coalition worked together to protect the sacred Bighorn Medicine Wheel. That group provided a vision of what could be accomplished with a coordinated and effective inter-Tribal coalition, but few sacred sites in Montana have the same level of inter-Tribal interest as the Bighorn Medicine Wheel. Yellowstone and Glacier are sacred spaces that Native people have celebrated and cherished forever, and their return to those places as cultural ambassadors and comanagers is the logical next step in bringing diversity, equity, and inclusion into our national parks.

Sunrise on Blue Lake, Crazy Mountains

Ancient Hunting Culture Sites

Defying Euro-American assumptions of nations living within defined borders, Native people in Montana moved around the region based on the availability of wild resources, especially bison. Although bison culture had reigned supreme across the Great Plains since the beginning of the Holocene, the heart of that way of life was in central Montana, where the largest and most resilient herds migrated north and south throughout the year. Like rivers of blood, these massive herds flowed across the land with their noses pointed into the blowing wind, prompted by gusts and guided by the smell of grass and water-filled valleys. The central Montana geographic feature known today as the Judith Gap marks a pinch point on the land where southward-migrating herds of bison were funneled into a bottleneck valley and then diverged into either a western path that led to the Paradise Valley, or an eastern route that directed them toward the Bighorn and Little Bighorn River valleys and then along the eastern front of the Bighorn Mountains. This eastern route, now a paved road known as "The Old Buffalo Trail," connects to the Yellowstone Valley in Laurel, just north of the confluence of the Clarks Fork and Yellowstone. The Nez Perce also followed this well-known trail through the Judith Gap on their flight to Canada in 1877.

The southwestward path that led bison from the Judith Gap to the Yellowstone Plateau straddled

Chief Mountain, autumn sunrise

the Crazy Mountains and went directly through the wind tunnel known today as the Paradise Valley. As these herds seasonally made their way back and forth to and from the Yellowstone high country, they would pass through buffalo jump sites marked by long lines of large, deeply embedded stones. Decades of research and careful documentation by local archaeologists have revealed over 60 miles of rock lines concentrated in a hunting zone about 20 miles long and 10 miles wide. These rock drive lines were used as fences to herd galloping bison toward a nearby cliff, where they would be wounded and subsequently harvested en masse. The rock lines were effective as fences because of the poor eyesight of bison and their basic instinct to stay within a safe space and avoid stepping onto uneven ground. Drive lines are commonly found at many buffalo jump sites in Montana, but the large-scale landscape engineering constructed by Indigenous hunters in the Paradise Valley displays an unprecedented level of permanent infrastructure for hunter-gatherer-traders. Within a global context, only some 8,000-year-old gazelle-hunting sites in modern-day Syria can compare.

The miles of aligned and embedded rocks also align with Apsáalooke (Crow) oral traditions and show a gradual sophistication in how Native hunters utilized the buffalo jumps. Although it's difficult to accurately date the lines, it appears that the oldest ones directed groups of galloping bison directly over the cliff, causing mortal injuries to the lead cow and residual social disruption and trauma to the entire group. The most recent rock lines were made within the last 200–300 years and mark a path that runs parallel to the edge of the cliff, rather than directly over it. As local archaeologists puzzled over these rock alignments, Apsáalooke (Crow) oral historians recognized them as engineered avenues used to preserve the life of the lead cow and still cull a sustainable harvest. During a site visit to the lines in the fall of 2022, Apsáalooke Tribal cultural experts informed researchers and other oral historians that their ancestors understood the value of keeping the lead cow alive when hunting bison, so the parallel rock lines protected that leader from falling over the cliff. Yet the lines' proximity to the cliff face and the large number of animals running alongside it inevitably led to many bison near the end of the herd drifting over the line and off the cliff, creating an efficient, practical, and respectful way to hunt that sacred animal.

Crazy Mountains atop central Montana prairies

Making Meaning for the Future from Tribal History

Montana Tribal communities have sought to recover from tremendous loss of land, life, and culture that marked their history in the nineteenth and twentieth centuries. A heavy burden of historical trauma has been inherited by contemporary generations of Native peoples here in Montana, and this ongoing challenge takes the most prominent attention of the Tribal governments, leaving them fewer resources and less capacity for working on public land issues outside the reservations. Yet Tribes continue to engage in more efforts off their reservations because they have begun to see the value and importance of working with partners to conserve and protect sacred landscapes.

The same is true on Tribal lands, where dozens of efforts to restore the natural ecosystems across Indian Country are in full swing today.

The future of land conservation in Montana is tied to Tribal communities in unique ways that empower collaborative efforts as well as Tribal sovereignty, and the many projects and ongoing partnerships throughout Indian Country speak to that truth. As the combined pressures from climate change, population increases, and continued development threaten fragile ecosystems, the successful working partnerships between Tribes, NGOs, federal agencies, and private landowners will become even more important to how our state responds.

There has been so much to learn during this long journey. For non-Indians, it began with ranching.

FURTHER READING

"The Genome of a Late Pleistocene Human from a Clovis Burial Site in Western Montana." February 12, 2014. *Nature* (42 coauthors including Shane Doyle). https://www.nature.com/articles/nature13025.

Indigenous Continent. 2022. Pekka Hämäläinen. Liveright.

"Native Americans Descend from Ancient Montana Boy." February 12, 2014. Michael Balter. *Science.*

Native Nexus. Shane Doyle's consulting firm. shanedoyle@yahoo.com.

Origin: A Genetic History of the Americas. 2022. Jennifer Raff. Twelve Books.

"Shane Doyle Links Montana Tribes, International Research into Prehistoric Boy." February 13, 2014. *Bozeman Daily Chronicle.* www.bozemandailychronicle.com

Six Hundred Generations. 2019. Carl M. Davis. Riverbend.

Making hay while the sun shines

CHAPTER 4

Ranching

Drive north over Monida Pass and roll down to a durable piece of Montana ranch country. The broad valley looks like Mongolia with the Beaverhead Mountains, Gravelly Range, and Snowcrest Range rising all around. This is a dry, cold place, but the Red Rock River provides precious water to herds of cattle and sheep. Mostly cattle, it seems; Hereford, Black Angus, and red, white, and black mixed breeds dot the fields like checkerboard pieces. Irrigated hayfields provide nutritious feed during cold times. When snow pushes elk down from the mountains, they take refuge in winter range on ranches. Everything looks in perfect order.

However, ranching in places like Beaverhead County is tough. Local families are not cattle barons as in the *Yellowstone* series—a fictional stab at Montana. They work hard for modest earnings, with most of their wealth tied up in property. Sometimes it doesn't pan out. In the century between the Civil War and President Kennedy, only three in nine ranch families held on to their land. Those places were settled by pioneer grandparents and then lost to bankruptcy or sold. Too much mortgaging, too little grass, low ag prices, and high out-migration. That led to a changing of the guard as new operators took over and adapted to natural and political changes.

One of those recent "operators" is media giant Rupert Murdoch, a character who *would* fit in *Yellowstone*. In 2021, Murdoch bought the 340,000-acre Beaverhead Ranch for $200 million. It takes 15 families to work the outfit, so the sale price had little to do with livestock. The amenity value—the natural beauty and biodiversity—elevated the price. The same is true all over the state; just look at ranch listings online. Prices are skyrocketing and the pressure to sell or subdivide ranches is what drives many land conservation projects today. A precious world is at stake; a natural/cultural landscape and honorable way of rural life.

Secretary of Agriculture Tom Vilsack understands. "Lots of farmers and ranchers are struggling. I get on my knees every day." But saving ranches is not just about prayers, it is about offering ranchers financial incentives for *not* developing their land, and finding a way to literally capitalize on safeguarding Nature. Upcoming chapters bring those stories to life.

Ranches hold Montana together economically, culturally, and ecologically. Some of the world's best beef, pork, wool, and wheat are produced here. Ranches also provide critical habitat for an array of wild species because public lands alone are not capable of sustaining them. Rivers and creeks flow through ranches supporting riparian and aquatic species. There are migratory birds, rare plants, and pronghorn. Without these habitats, Montana would be a poorer place.

Historical Roots

Ranching seems eternal, but it dates back only to the mid-nineteenth century in the state. Cattle are newcomers too, just like horses, sheep, goats, and pigs. These animals were domesticated in the Middle East and Eurasia 10,000 years ago and then spread. In England, Anglo grazers cooperated with Saxon feedlots to create an early form of ranching. Transhumance arose all over in response to seasons by moving cattle between summer and winter pastures. It was perfected in Spain, where dry valleys sit below wet meadows in the Pyrenees and Sierra Nevada, country like Montana in many ways. Cows were driven to proper range as needed and irrigation ditches were dug on the Moorish model straight from Morocco. The Spanish word *acequia* comes from Arabic *al sakiya*, "the ditch." Irrigation and cattle movement made year-round ranching tenable in diverse environments. All over Eurasia and Africa, animal husbandry modified cattle into breeds suited to local conditions. Instead of hunting for meat, humans now raised livestock. That breakthrough along with plant domestication led to massive food production and an increase in population free to invent new technologies, including ships and navigation.

In 1493, an Italian named Christopher Columbus brought the first cattle to the Americas. The Spanish settlers he carried spoke words we still use: lariat, chaps, vaquero (buckaroo), and rancho. In time, herds were driven north by Mexicans, Texans, and African Americans into California and the Great Plains. In Montana, beef was commercialized by settlers like Conrad Kohrs, a German. He bought land in the Deer Lodge Valley from a French Canadian gold miner named Johnny Grant and expanded it into a cattle kingdom covering 10 million acres—10 times the size of Glacier National Park. Meat was sold to gold miners, settlers, and railroad crews. Some fur traders bought burned-out cows from wagon trains, fattened them on free grass, and then resold the beasts. The Grant-Kohrs Ranch National Historic Site near Deer Lodge is a 1,600-acre reminder of those times.

In Eastern Montana, herds of half-wild cattle were driven north from Texas on the Bozeman Trail. Larry McMurtry's book *Lonesome Dove* is a fictionalized but striking account of that time. Montana *was* a kind of promised land, wetter than the Southern Plains and highly productive. Bison herds were nearly wiped out by the 1870s to feed miners and railroad crews. Hides and bones were also valuable commodities. Killing bison was also a military strategy to weaken Native Americans' resolve to fight since these huge animals formed the sacred and practical foundation of many tribes. With bison gone, cattle spread across land claimed by the US government through a trail of broken treaties with Indian nations. Vast homelands became open range where stirrup-high grass was free for the taking. Cattle drives into the newly declared public domain brought immense herds to the prairies. The land seemed empty, endless, and bountiful. In 1883, the Northern Pacific Railway reached Miles City and completed the second transcontinental line across America. Herds expanded since cattle could now be shipped east and west to booming markets. Indian lands contracted.

The open range was a cowboy's dream, but a fitful one. Rustlers made off with livestock, predators killed unwatched animals, and the prairie could be really dry. In *Montana Places*, the author said of early Miles City: "The railyard was a high-smelling warren of holding corrals, chutes, and loading docks filled with bawling cows. Downtown was a clutter of saloons, boarding houses, liveries, bordellos, and banks. Cowboys, clerks, and cavalrymen kept things lively."

The "open range" system wasn't land stewardship; it was a go-for-broke push to graze as many cows as you could for free. Calves were branded in the spring, set loose for a year, and then rounded up and sold. Herds mixed, so brands were crucial. The Plains were a "commons" with no one in charge—got cows, graze them.

Frontier cowboys embody one of our country's most enduring myths—loyal to their companions, fiercely independent, with a high code of honor. They were seen as paladins of the prairie, noble and pure, protecting livestock in all weather, shielding the innocent from harm. Some of that is true, but cowboys were regular people, mostly young men, uneducated migrant workers who labored hard for low wages. White, Black, Latino, and Native American. Their job was to drive herds through rain, hail, snow, mud, dust, and sun for months at a time. A saddlesore trip through paradise.

By 1885, the Montana Territory had over 500,000 cows roaming the Great Plains. In time, the prairies became overgrazed, with some of the worst of it around Miles City. In 1886, winter temperatures hit 30 below and cattle wandered into town desperate for food. The next summer, temperatures soared to over 100 degrees and range fires burned up what grass was left. Fall gales covered the stubble with ice.

Then it got real.

In *Montana Places*, the author described the calamity this way: "In January of 1887, a month the Northern Cheyenne call 'the Moon of Cold Exploding Trees,' temperatures dropped to 40 below and stayed there. Blizzards covered the beaten landscape with snow higher than a cow's belly.

Herds of cattle stumbled around, snow-blind and doomed. They sought refuge in coulees where they huddled together and froze to death. The grotesque scale of the losses was revealed the next spring. The prairie was strewn with carcasses and wolf-bit bones. In the winter of 1886–87, 362,000 cows died of starvation and exposure; more than 70% of all cattle in the Territory. In places, it was 90%."

The open range system failed.

The Modern Ranch

Ranchers responded to that collapse by buying land and building headquarters—the historical heart of today's family outfits. They strung barbed wire fences to create pastures, fed cattle over the winter, and developed year-round water sources. Salt blocks were laid out and animal husbandry replaced indifference. Keeping a close eye on range quality became essential.

Modern ranching was born when people embraced land stewardship.

"The ranch" is an adaptive strategy for making a living and caring for the Earth. It is a diverse practice in many ways, but always focused on an ancient task—profitably raising domesticated animals in ecologically sensitive country. Access to grass has always been tenuous, especially for ranchers holding public land grazing permits who knock heads with their landlord—federal or state government. Conflicts over the animal unit months (AUMs) of grazing allowed, fencing requirements, and other regulations mark that uneasy relationship. Leased ground is an essential part of a ranch's viability and value. Real estate listings today distinguish between "deeded acres" and leased ground, but the assumption is that the two go together. Mostly that's true, but uncertainty creates stress within ag families. Ranchers who have no grazing permits must own more land, which creates a debt load for some or a family inheritance for others that must be protected from estate taxes.

Faced with all this, ranchers tend to be conservative, but they don't necessarily hold the same beliefs. They are individualists. A "home ranch" provides a strong geographic identity, income, way of life, and haven. Rural unease, wariness of change, and mistrust of outsiders is the result of 170 years of challenges ranchers have dealt with, such as environmental critics.

"No Moo in '92." "Cattle Free by '93." These slogans represented a movement aimed at "removing livestock from public lands to conserve native biodiversity," to quote the subtitle of Debra Donahue's book *The Western Range Revisited*, a lengthy deposition to stop all grazing in the public domain. She argues that Bureau of Land Management leases produce a tiny share of US livestock products and that the program costs too much to administer. She believes that cattle and sheep cause widespread ecological damage. In *Welfare Ranching*, George Wuerthner complains bitterly that grazing permits are "subsidizing the destruction of the American West." While localized impacts on creeks and grasslands exist, deep experience out there disproves those alarmist accusations of ruin.

Wendell Berry wrote, "The longstanding conflict between ranchers and conservationists is not only pointless, but ruinous for both." Courtney White agreed in *Revolution on the Range*. The scary title belies his conciliatory point—turning conflict and suspicion into respect, cooperation, and trust. White foresaw recent partnerships between agricultural organizations, agencies, NGOs, and most importantly, landowners. Mutually beneficial ways emerged to work things out.

Today, most of us are sensible when it comes to ranching. That political position is so different it's been called the "radical center." To get there requires some basic understanding. Most ranch families are not income rich and often need outside jobs to make ends meet. Land is their main asset. Sons and daughters may not want to ranch—then what? Trespassing can be a pain during hunting season. Ag prices are volatile and based on regional and global market forces. Regulations can seem random and out of step with a rancher's needs. Operators also face challenges in balancing livestock and wildlife on their places, but they do it anyway. As a result, ranchers feel baffled when they are seen in a negative light.

Ranching is more of a cultural legacy than a cash-based choice. In *Let the Cowboy Ride*, geographer Paul Starrs emphasizes the growing range of compensations ranchers benefit from: Conservation Reserve payments, quiet,

ecotourism, prestige, feeding people, and enjoying a meaningful life. Ranchers are not ransackers, and Aldo Leopold's land ethic is widely embodied in their everyday lives. To do otherwise brings economic and ethical harm.

Today, ranches and farms cover 62 percent of Montana's private land, where 27,000 families act as stewards mostly free of charge. Much of the state's wildlife habitat and scenic open space is found on private lands. If they disappear, Montana is degraded. Beef, lamb, pork, and wheat are strong currencies in the state economy, worth $5 billion per year. Ranches are the natural foundation of commerce and conservation—a resource to respect, the focus of relationships to cherish.

After the open range period, ranching became professional, supported by science, improved breeds, soil conservation districts, grazing districts, and federal programs that reward strong stewardship practices. The Conservation Reserve Program and Wetland Reserve Program provide cash in exchange for conservation. In some Eastern Montana counties, these payments help keep ranching economically viable. Montana now has one million beef cattle, in seventh place nationally. For perspective, Texas has five million cows. Beyond numbers, ranching is now the most efficient and sustainable way of caring for vast areas of Montana. That is why so much attention is placed on keeping ranches intact.

But ranches come in all sizes and expressions.

The Happy Hollow Ranch in Mineral County is small, a modest base of deeded acres. Toots and Wayne Bricker have run the place for years on a rolling patch of ground in Tarkio, a wide spot east of Superior. The Happy Hollow is a homemade operation put together with no expectation of wealth. It's a multigenerational home. Toots worked in the Mineral County Treasurer's Office for years, earning outside income that helped back on the ranch. Wayne runs the cattle operation and, along with Toots, grows wheat with machinery and a sense of humor. USDA grain subsidies help with cash flow. The Happy Hollow is a family-run outfit, much smaller than the 2,100-acre average size in the state. With low subdivision pressure, the ranch may survive if a next generation can be found to run it.

OPPOSITE: Harvest time in ranch country

The Valley Springs Ranch covers a thousand acres next to the Bitterroot River. It's gently used and mostly left alone. Musician Huey Lewis bought the place as a retreat from the road and a base for fishing trips. He had his stretch of Mitchell Slough restored by an aquatic biologist named Dave Odell—a fish whisperer. Trout now thrive in a former channel and osprey are grateful. Huey donated a conservation easement on the place some years ago to keep the ranch from ever being subdivided. In the booming Bitterroot Valley, that was a real contribution. Conservation easements are a voluntary tool you will hear much more about in upcoming chapters.

The IX Ranch covers 120,000 acres around the Bears Paw Mountains in the Glaciated Great Plains ecoregion. Big Sandy is the trade town around here, and cattle are still worked on horseback. The owners call it "a model multigenerational Montana ranch." Stephen A. Ross is the boss, a man who served as president of the Montana Stockgrowers Association. IX literature says, "Enlightened utilization and conservation of natural resources and superior economic returns are the primary benchmarks of accomplishment." Cattle are tended well. "We care for them as we would our own family members to ensure their time spent with us is as enjoyable as possible." Heartfelt words that seem odd, until we remember we all kill to eat. It's a major operation, and each year the IX Ranch sells about 1,300 steer calves, 800 yearlings, and 400 cows. Bulls are bought in Sidney, Hobson, and Billings, Montana, from a network of trusted breeders. IX Ranch principles are focused on "integrity, respect, and superior stewardship." Many ranchers may feel the same, but this outfit says it out loud.

Regardless of politics, ranchers across Montana are banding together to create cooperative visions for the future. Rather than be steamrolled by change, they are taking control. The Madison Valley Ranchlands Group's mission is to improve the economic vitality of family operations while sustaining grazing and wildlife habitats. The Big Hole Watershed Committee focuses on meeting the needs of ranchers, sportsmen, outfitters, residents, and the land. The Nature Conservancy and Trout Unlimited are active here, focusing on providing water and habitat for fluvial Arctic grayling. The Blackfoot Challenge is a working group with astonishing results in land conservation, as you'll see in chapter 11.

However, Montana rangeland has declined by 1.3 million acres since 1982, and cropland has dropped by 1.7 million acres. That's only 5 percent of the state's agricultural land, but most of that loss was caused by land subdivision developments in compressed Rocky Mountain valleys around Missoula, Bozeman, Kalispell, Helena, and Hamilton. The remote Great Plains ecoregions are little impacted by growth. In fact, many counties are declining in population. The choice out there is about plowing prairies to create wheat fields.

Ranching in Montana has been buffeted by challenges since it began. Yet all over the state, ranches still bind the landscape together. Montana operators have survived hard times and created a way of life we all benefit from. But what is the legacy of those who attempted a much different form of land stewardship?

Homesteaders.

FURTHER READING

The Cowboy Way: Seasons on a Montana Ranch. 2000. David McCumber. Mariner Books.

Let the Cowboy Ride: Cattle Ranching in the American West. 1998. Paul F. Starrs. Johns Hopkins University Press.

Lonesome Dove. 2010. Larry McMurtry. Simon and Schuster.

Montana Places: Exploring Big Sky Country. 2000. John B. Wright. University of Minnesota Press.

North American Cattle-Ranching Frontiers: Origins, Diffusion, and Differentiation. 1993. Terry G. Jordan. University of New Mexico Press.

Sketches from the Ranch: A Montana Memoir. 2008. Dan Aadland. Bison Books.

Crazy Mountains in the distance

CHAPTER 5

Homesteading

As I looked across the rolling expanse of prairie, filled with the beauty of a Montana sunset, I sent up a little prayer of thanksgiving from my heart for this our very first home. Only a rectangle of sod, raw and untouched by the hands of man, but to us it was a kingdom.—PEARL PRICE ROBERTSON, a Big Sandy homesteader, 1911

I have stood in the doorway of our shack, with my heart full of sadness and loneliness and listened to the wind. It is an incessant screeching, whining and screaming wind, and it seems to be heard nowhere except in Montana.—SUE HOWELLS, a Choteau County homesteader, 1918

Montana is a geographic Rorschach test. Based on our personalities, we see it has a haven of transcendent wonder or a harsh terrain of exile. Before the Indian Wars were over, Congress wrote a slew of laws designed to settle Native land in the West by creating farms. Manifest Destiny insisted that the United States would be a continental country regardless of the brutalities required to take the land.

The idea was blunt: it is manifestly obvious that God approves of White people and western commerce. Indians are in the way. When President Lincoln signed the Homestead Act in 1862, he said, with no sense of irony, "This will do something for the little fellow." America tried to prove it was not elitist like Great Britain. Our country would have no royal holdings; we were a society based on the right of every citizen to own land. But this fine commitment was aimed largely at White people at the expense of Indigenous ones.

The first 160-acre homestead in Montana was claimed by a man named David Carpenter. It was in the northwest quarter of Section 18, Township 10 North, Range 3 West—a modest spot near Helena. That description came from the 1785 Rectangular Survey System, which laid a grid of lines across the frontier in order to give land away or sell it. Each block of 36 square miles could be divided further based on an aliquot parts description. The system was geometrically precise and repeatable across immense landscapes. A "section" is a square mile, 640 acres. A quarter section is 160 acres. The US General Land Office hoped that clearly described land titles would make settlement simple. They didn't. The seemingly egalitarian Homestead Act led to lots of shenanigans. A loophole called the "commutation clause" allowed homesteaders to quit after six months and sell the land to a speculator. Resale values grew as monopolies gained control.

In the nineteenth century, Montana was little known to Euro-Americans. President Jefferson's acquisition of the Louisiana Purchase happened before the Lewis and Clark Expedition. Those explorer accounts provided basic geographic information about Montana but left so much unrevealed—especially facts about climate, soils, and water, the essentials of agriculture—and they largely ignored the knowledge of Indian people. All homesteading bills that followed were empowered by a geopolitical drive to settle the West even without a Northwest Passage to the Pacific Ocean.

President Jefferson and Frenchman Michel Crèvecoeur were elites, but they championed the idea of "yeoman farmers." Agriculture was seen as a morally superior way of life. To them, the American West should become a "fee simple empire" based on individual rights in beautiful country. As a result, the region (including Montana) attracted people from California, the Midwest, and back East. Same as it ever was. In *Virgin Land*, historian Henry Nash Smith wrote, "The image of a vast and constantly growing agrarian society in the interior of the continent became one of the dominant symbols of 19th century American society—a collective representation, a poetic idea that defined the promise of American life."

Geographic myths burdened homesteading from the start.

In *Commerce of the Prairies*, Josiah Gregg promoted the biggest misunderstanding of all: "rain follows the plow." He wrote, "Why may we not suppose that the general influences of civilization—that extensive cultivation—might contribute to the multiplication of showers?" Absurd as this sounds, it was widely believed in the humid East, especially by politicians pushing for western expansion by regular people seeking a fresh start. The 1862 Homestead Act was based on a climatic falsehood, but back then, there was no scientific consensus about it.

John Wesley Powell's *Report on the Lands of the Arid Region of the United States* was drawn from his arduous explorations of the West. He trudged across prairies, climbed mountains, floated the Colorado River, and collected climate data from US Cavalry forts. The title and the first paragraph of Powell's report to Congress seemed to resolve the dispute. "The eastern portion of the United States is supplied with abundant rainfall for agricultural purposes . . . but westward the amount of aqueous precipitation diminishes until at last a region is reached where the climate is so arid that agriculture is not successful without irrigation."

Others disagreed. Dr. Ferdinand Hayden was an explorer-scientist who promoted another damaging myth: "It is believed that the planting of ten to fifteen acres of trees on each quarter section (160 acres) will have a most important effect on climate, equalizing and increasing moisture and adding to the fertility of soil." In *Montana Places*, the author wrote, "If the yeoman's holy plow couldn't civilize aridity maybe sacred trees could." Even Bernard Fernow believed it, and he was the first president of the Society of American Foresters. Regular people ate it up.

In 1873, Congress passed the Timber Culture Act, which offered 160 acres free to any settler who promised to plant 40 acres into trees. That dreamy provision was enthusiastically evaded. Settlers were paid to "relinquish" their claims by companies that accumulated large holdings. The commissioner of the General Land Office said, "I can truthfully say that I do not believe one timber culture filing in a hundred is made in good faith." The Timber Culture Act was soon repealed, but nearly 10 million acres had been distributed.

The Desert Land Act of 1877 was a grudging nod to aridity. Homesteaders could buy a section for $1.25 per acre *if* they applied irrigation water and "proved it." For land next to rivers, this worked, but not all Montana claims were hydrologically blessed. John Wesley Powell believed that 640-acre homesteads were much too small. He recommended four-square-mile homesteads (2,560 acres) per family *if* they had irrigable land next to a river. Powell believed that western state boundaries should be based on watersheds, since water supply was the foundation of life. His wise notion was laughed off in the halls of Congress.

By the 1880s, railroads had brought rapid settlement to Montana. The "Empire Builder" and other trains connected us with both coasts. James Hill, head of Great Northern, bragged, "Give me enough Swedes and whiskey and I'll build a railroad through Hell."

In 1909, Congress passed the Enlarged Homestead Act, which played a transformative role in the settlement of Montana. It promised 320 acres of nonirrigable land free of charge, a way to settle without paying a railroad company for the privilege. The "summer fallow system" of growing wheat made homesteading plausible in Montana's semiarid plains. Prairies were plowed under and wheat was planted in one strip, with the adjacent strip left fallow for a year to accumulate moisture. It worked.

Waves of settlers arrived in the "Hi-Line" of Montana in railroad towns like Glasgow, Malta, Havre, Chester, and Shelby. Great Northern's ads blared, "Millions and Millions of Acres for Sale in Montana at the Lowest Prices Ever Offered—$2.60–$4.00 per acre! The Best Homes for 10,000,000 people! They Will All Become Prosperous!" Broke people got off a train, claimed a 320-acre piece, and were dumped by a "Locator" in the middle of nowhere, left to their own devices. In southern Montana, people bought land directly from the Northern Pacific, which had been granted 13.5 million acres along its right-of-way by the US government. The company sold 11 million acres in eight years and optimism ran high. A Northern Pacific promotional poster showed a farmer turning over prairie sod and gold coins pouring out. The message was clear: fertile wealth is there for the taking.

Grain elevator near Laredo, Montana

Rainfall was excellent in Montana beginning in 1910, and yields exploded from 11 million bushels to 42 million in the "miracle year" of 1915. People called it the Bonanza Wheat Boom. In *Montana Places*, the author wrote, "Prices for Montana's superb, high protein spring and winter wheat reached record levels during World War I (with competing markets damaged). Homesteaders took all this in stride. The promoters promised great things, and it all was coming true." Maybe rain *did* follow the plow. Many homesteaders abandoned the summer fallow system and planted wheat every year. The price was too good to resist, and the future seemed bright. The federal government urged homesteaders to do their patriotic duty and spend wheat profits on more land and machinery. World War I was raging, after all. The slogan was "Food Will Win the War!"

When the Enlarged Homestead Act was passed in 1909, Montana had 360,000 residents. By 1918, the population had grown to more than half a million—a 40 percent increase. New towns rose everywhere, and new counties were formed to absorb the growth. Problem was, rainy weather was mistaken for a wetter climate.

Soon, normal aridity returned and soil turned to dust. Mobs of grasshoppers competed with desperate farmers for puny crops, and Russian thistle took over. Ceaseless winds blew soil off parched fields creating an early Dust Bowl, and at the end of World War I, wheat prices crashed. Homestead life turned psychologically grim and financially disastrous. Many settlers lost their land to foreclosure or sold to ranchers, so that by 1920 more than 200,000 settlers had left the state. The Hi-Line out-migration was striking, with 65,000 of 85,000 homesteaders quitting farm life. Land prices collapsed and banks went bust. Historian Derek Strahn wrote, "By 1925, more than 2 million acres were abandoned and half of Montana's farmers lost their land." The Great Depression and relentless drought finished off many others.

Small homesteads had mostly failed.

Adaptation

Of 32 million acres of homesteads in the West, only a quarter survived. The rest were acquired by ranchers or reverted to the federal government, much of that becoming part of the Bureau of Land Management.

In Western Montana, small homesteads succeeded beside rivers. Large farms worked in the Great Plains if properly managed using the summer fallow system or irrigating. Today, most grain outfits in Montana cover thousands of acres, patching together lots of former homesteads. Some became mixed cattle and wheat outfits. The township and range survey system is still revealed in the gridded pattern of roads and fence lines. Wheat fields are now planted north-south to reduce erosion from westerly winds.

You can see this on Google Maps. Look at the US-Canada border north of farmlands by Chester and Havre. Alberta and Saskatchewan still have vast expanses of unplowed prairie. The border shows a stunning contrast in land stewardship history. The Canadian homestead act was called the Dominion Lands Act, and many areas were never plowed. A scientist named John Palliser said it was too dry and he was right.

In Montana, abandoned houses and barns are scattered across the Great Plains. Every structure reveals a family dream—an honorable urge to make their own way. There is an epic in every property, personal stories we will never know. Most homesteaders left the state broke and crushed, but some stuck it out. You can talk to their relatives today across Montana—Scandinavians and Germans mostly, but settlers came from everywhere. They are the descendants of "stickers," tough people who managed to stay.

People like Nordie, raised on a homestead near Nashua, north of Fort Peck Reservoir. She is kind and no-nonsense, raising money for the Sons of Norway by selling Vikings at the Missoula County Fair—batter-fried meat on a stick. Nordie is affectionately called "The Nashua Flash." Then there's Rugga, who still manages the family cattle operation near Two Dot. He knows everyone in the state and has stories to match. Some of them are even true. "Work, church, and home," he'll say at the Two Dot Bar, and then take a sip of beer as a punch line. In case you wondered, he'll point at a sticker on the wall: "We don't call 911." Families still protect each other out here.

Frontier life on the prairie has been highly romanticized. Novels by Zane Grey, Hamlin Garland, and Laura Ingalls Wilder paint it in old colors. Today, Deirdre McNamer's books weave

together new insights about Montana without dropping a stitch. She was born and raised in Conrad and Cut Bank, so her stories rise up from the ground. McNamer's novel *One Sweet Quarrel* covers the homesteading land game better than any history. The plot is striking and the prose is strong. In 1923, Shelby, Montana, tried to promote itself by hosting the Dempsey-Gibbons heavyweight championship fight. That gamble bankrupted the place. Shelby is now a small railroad and wheat community on the Hi-Line, part of the productive Golden Triangle. As in many homesteading regions, soil and grain became lasting sources of honorable prosperity—a way to hold Montana landscapes together, and an enduring reminder of limits and realities on the prairie.

That is the land stewardship lesson of homesteading in Montana.

FURTHER READING

Commerce of the Prairies. 1967. Josiah Gregg. Bison Books.

The Great Plains. 1931. Walter Prescott Webb. Ginn.

Montana: An Uncommon Land. 1959. K. Ross Toole. University of Oklahoma Press.

The Montana Medicine Show's Genuine Montana History. 2014. B. Derek Strahn. Riverbend.

Montana Places. 2000. John B. Wright. University of Minnesota Press.

Montana's Homestead Era. 1987. Montana Geographic Series. Daniel N. Vichorek. Random House.

One Sweet Quarrel. 2013. Deirdre McNamer. Amazon /Encore.

Report on the Lands of the Arid Regions of the United States with a More Detailed Account of the Land of Utah with Maps. 1879. John Wesley Powell. https://pubs.usgs.gov/publication/70039240.

Virgin Land: The American West as Symbol and Myth. 2007. Henry Nash Smith. Harvard University Press.

The city meets the country

CHAPTER 6

Land Subdivision

"Why come to Montana? Some move here to get away from drought and floods, some reap the fortunes realized by the rise in value of land in other states. Many seek a less-crowded environment."

That was written in 1909. The Montana Department of Publicity was prophetic. This pamphlet came a century after Lewis and Clark and just 68 years after the first White settlement in the state. It was called St. Mary's Mission, in the heart of the Bitterroot Valley where a log Catholic church was built at the request of the Salish Nation. They foresaw the massive change coming before it crested the horizon.

The Bitterroot

The first valley settled in Montana is now one of the most subdivided. It promises a fresh start, a safe place with rugged peaks and plenty of trout in the river. A neo-West paradise.

But to the Salish, the Bitterroot is a lost homeland, a place so fertile that White settlers arrived in the nineteenth century without permission—taking, farming, excluding. St. Mary's Mission became the town of Stevensville, named for George Stevens, a territorial governor born in Massachusetts. By 1855, the Salish were being overrun. They signed the Hellgate Treaty and then tried to hold on. However, by the 1870s, most of the nation was driven north to the newly formed Flathead Reservation, with armed soldiers watching every move. Chief Charlo and his band of 342 people held out until 1891, when suffering forced them to join their cousins in St. Ignatius. They were promised a fine church, land, and respect.

Only part of that came true.

Heading north, Charlo's band crossed the Higgins Avenue Bridge over the Clark Fork River in Missoula. A rainy trudge of exhausted refugees, a forced removal from a homeland.

With the Salish gone, land speculation thrived in the Bitterroot. In 1910, a group of businessmen from Chicago built a 60-mile-long irrigation canal from Lake Como north to Florence: the "Big Ditch." Over 14,000 acres were split into 10-acre lots and sold for huge prices at the time, up to $1,000 per acre. Each parcel was marketed as an "apple orchard tract." Frank Lloyd Wright designed the Bitter Root Inn, where prospective buyers were softened up with thick steaks, French wine, and rounds of golf.

Soon, 100,000 acres were subdivided, sold, and planted into McIntosh apple trees. Early yields were immense, but the whole scheme failed. There wasn't enough water, soils were thin, and Montana could not compete with fruit growers in the Pacific Northwest. Besides, the "apple farmers" were city people with no experience on the land.

By 1918, the whole enterprise went bankrupt. Swindlers took apples and never paid, and then codling moths turned the fruit into mush. Investors suffered immense losses and fled. In time the apple trees rotted or were cut down for firewood, and the Bitter Root Inn burned to the ground, leaving no trace. But the dense pattern of land subdivision was set in Ravalli County.

Subdivisions and Planning

In the early 1970s, another land subdivision boom came to the Bitterroot and other parts of Montana. Residential development spread with no consideration of its effects on ranching, wildlife habitat, open space, and the public expense of road maintenance, schools, fire protection, and police. Ranchers and farmers who needed cash simply split their ground into lots to keep their operations afloat. Absent economic alternatives, who could blame them? But what about the future of the Montana landscape?

In response, the state wrote a visionary new constitution in 1972. It contained a bold provision: "The State and each person *shall* maintain and improve a clean and healthful environment in Montana for present and future generations." More on this later.

In 1973, the legislature passed the Montana Subdivision and Platting Act. This law gave city and county governments the right to "promote the public health, safety, and general welfare by regulating the subdivision of land and prevent overcrowding." It required "development in harmony with the natural environment." The act authorized public reviews and local approval of plats, so planning boards were formed across the state to evaluate new developments. Surely now, rural subdivisions would be managed. It never happened.

A geographer named Mark Beardslee showed why. He scoured land surveys filed in nine fast-growing counties and reported that "of the nearly 130,000 acres divided between 1973 and 1976, about 120,000 acres, 93% of the total, were divided without public review." The Montana Environmental Information Center reported the same outcome.

How could this be?

Exemptions to the Subdivision and Platting Act outnumbered cases in which the law could be applied. The following types of land splits *could not* be reviewed by local governments but *could* be legally recorded as Certificates of Survey:

1. 20-acre parcels and larger (surveys created swaths of 20.01-acre lots)
2. "Occasional Sales" of one parcel per year (a chain of buyers could create 20 one-acre lots in the same year)
3. "Family Transfers" of one parcel per year (a chain of family members could quickly divide a parcel without review)

The Montana subdivision law turned out to be a de facto homestead act. In Ravalli County, 11,644 acres were divided using exemptions during the first three years. Only 103 acres were reviewed as "subdivisions." Unregulated rural development continued across Montana under the banner of law.

Between 1970 and 2020, Ravalli County grew from 14,000 residents to 48,000, largely from in-migration. About 1,500 new people arrive in the

Bitterroot each year. The pattern and density of subdivision is striking. Houses and businesses now parallel Highway 93 for 50 miles from Missoula south to Hamilton. Gas stations, log home companies, diners, gravel pits, and retail shops dominate where ranches once existed. There is nothing inherently wrong with the decisions of these landowners. They have families to support and bills to pay. Besides, new people need a place to live.

But historical ranches and farms are being rapidly subdivided to accommodate the rising demand. Cattle can't compete with the prices new people are willing to pay for residential land. The Bitterroot has more jobs now, but agriculture, rural life, open space, wildlife habitats, and natural beauty fade with every survey line. The story is the same around Missoula, Bozeman, Kalispell, Helena, Billings, and other Montana places. In the last half century, the state has been transformed from an isolated mystery into a national destination.

Montana had 694,000 residents in 1970; by 2020 it had 1.1 million. The state is on pace to add another 100,000 citizens in the next decade. With the mountains held mostly by public agencies, development is confined to the Montana Valley and Foothill Prairies. As a result, real estate and construction are now significant industries, with just 15 percent of Montana's GDP coming from agriculture, forestry, and mining.

However, this is deceptive. Ranches and farms generate immense societal benefits as land uses. Agricultural operators manage wildlife habitats, watersheds, and scenic vistas, usually at no cost to the public. These activities are called "ecological services." Rural culture and customs are also essential for Montana to remain Big Sky Country. Therefore, the true value of agriculture cannot be reduced to earnings and employment figures, so the impacts of subdividing these private lands cannot be brushed aside.

The rapid expansion of development in Montana continues to happen legally, in compliance with state law. Comprehensive plans are written by cities and counties to guide growth. Planned Unit Developments (clustered houses), setbacks, and design standards can improve the functionality of subdivisions. Obvious land-use conflicts are minimized and traffic flow is planned. Zoning and special tax districts help create livable cities with vibrant downtowns. Land-use planning has virtues, but in rural valleys its application is still very basic. As long as the septic tanks work and the roads are up to country specs, it's all good.

Regulatory land-use planning is not designed to conserve places. Despite lofty words in Comprehensive Plans about quality of life, wildlife habitat, and scenic integrity, most developments are approved. This happens for a simple reason.

Regulations determine how land *will be* developed. They have nothing to do with protecting significant areas. People troubled by widespread development of private land have a lot to learn about real estate, economics, culture, and the US Constitution.

The Fifth Amendment states, "Nor shall private property be taken for public use without just compensation." Governments can condemn land to build roads and power line corridors—then they pay the appraised value for it. But the development of privately owned ranches and farms *cannot* be prevented. Landowners must receive "just compensation" for their economic loss if their right to develop their land has been confiscated. No jurisdiction could afford it or even try.

This is because of the "Takings Clause" of our Constitution.

The case of *Lucas v. South Carolina Coastal Council* reveals why using "regulatory takings" to protect Nature should be rejected as unethical and futile. A man named Lucas bought oceanfront land to build two houses on, but the coastal council blocked his building permits for environmental reasons. Lucas sued and the case made it all the way to the US Supreme Court, which ruled in his favor in 1992. The State of South Carolina was ordered to pay him $850,000 for his property. And here comes the irony—the state then sold half the land to raise the needed cash. Today a 5,000-square-foot house stands on that shoreside ecosystem. Everyone lost, including the natural world.

It turns out you can't take property rights away from people. Those rights have real economic value and people disapprove of being robbed. Montana's subdivision law remains a clear victory for private property rights, but it creates deep problems as land policy. Taken to its logical conclusion, it will cause much of Montana to run out of ranches, wildlife habitats, and natural beauty.

People feel depressed and powerless to alter that outcome. Phrases have arisen in Montana to grapple with the pain. "The Last Best Place" has both pride and defeat embedded in every letter. Norman Maclean's *A River Runs through It* rings like a literary bell, but the Blackfoot was so altered by logging and mining that it could not play itself in the movie. Historian Joseph Kinsey Howard figured that Montana was "High, Wide, and Handsome" even while residents grappled with the complex legacy of copper mining. And we still are, using the Superfund program to clean up land and water as best we can.

Bud Guthrie's book title *The Big Sky* comes the closest. While Montana's sky may not be physically bigger than Canada's, the phrase evokes immense opportunities and responsibilities. We have the space and freedom to find ways of respecting each other and this astonishing landscape. As Joseph Kinsey Howard wrote, "The sunset brings infinite promise." Native people might disagree.

Indian reservations were Montana's first taking and major land subdivision. The Fort Laramie Treaty of 1851 stripped most of the state from Native hands. In 1855, the Flathead and Blackfeet treaties took even more. Current reservations account for only 9 percent of Montana (8,366,000 acres). The other 91 percent was taken by force and a trail of broken words. Land was even subdivided within reservation boundaries.

The Dawes Act of 1887 was officially called "The Allotment of Lands in Severalty to Indians on the Various Reservations." This federal bill was designed to sever Native people from communal identities. Land allotments (ownerships) broke up tribal holdings into parcels that could be sold. The message was clear: abandon traditional ways of life, become farmers, and adopt White ways. Then in 1910, the federal government opened up land inside the Flathead Reservation to homesteading. Today, only 60 percent is in tribal hands, with land development spreading across the rest.

In all settings, regulations are ill equipped to accomplish land conservation. But what is the alternative to regulatory tools? What lies beyond the constitutional "takings" line in regulating private property?

Land Conservation Emerges

What lies beyond the "takings" line is giving away or selling your right to develop the land. And doing this voluntarily while receiving financial compensation. The next chapter explores this in detail, but here's an introduction. In conservation easement deals, development rights are donated to a qualified land trust or unit of government in exchange for income and estate tax breaks. Or the landowners sell the easement. In both cases, they still own the property and use it as they did before. Key land can also be bought for conservation or traded for less ecologically important property. Conservation real estate practices arose in the wide gap left by land-use regulations. In Helena, Bozeman, Missoula, and other cities, open space systems are created when land is purchased, trails built, and development avoided. Planners and citizens finally learned the limits of regulation. Nonprofit groups like The Nature Conservancy, The Conservation Fund, Montana land trusts, and government agencies now protect more and more land from development.

In chapter 8, the statewide land status map shows the general pattern. Conservation easements exist on some 20,000 acres of ranches in the Bitterroot Valley, in Sula, Conner, Bell Crossing, and north to the historical Maclay Ranch. The Bitter Root Land Trust worked with landowners to create a 10-mile corridor of conserved ranchland in the Burnt Fork extending east from Stevensville to the Sapphire Mountains. This group received the Land Trust Excellence Award from the Land Trust Alliance.

The privately owned Teller Wildlife Refuge covers 1,200 acres beside the Bitterroot River. Montana Fish, Wildlife and Parks created a string of fishing access sites such as Poker Joe, Chief Looking Glass, and Bell Crossing. The Lee Metcalf National Wildlife Refuge covers 2,800 acres of riparian and wetland habitats. It is named for US senator Lee Metcalf, who grew up in Stevensville. The refuge supports some 242 bird species—including bald eagles—40 mammals, many trout, and 11 species of reptiles and amphibians. A rookery of great blue herons is home to 15 nests. A grizzly bear recently denned beside the Bitterroot River. The Selway-Bitterroot Wilderness secures 1.3 million acres of spectacular glaciated mountains, watersheds, and remote wildlife habitats. Here, visitors experience a bedrock American value, true freedom—terrain where you can die for being stupid or unlucky. The wilderness was created in 1964, as was the Anaconda Pintler Wilderness, which is the source of the East Fork of the Bitterroot River.

The Bitterroot is both extensively developed and somewhat conserved.

How do Salish people view all this? As we talk about "land conservation" in the Bitterroot and elsewhere, Native American losses remain mostly unresolved, including water rights issues. Today, practical opportunities exist to redress those grievances. However, new management partnerships require that we see Montana history not as the distant past, but as a continuing truth.

Go to the Bitterroot Valley today, specifically to the Lee Metcalf National Wildlife Refuge. Signs call this "Salish land," honoring their cultural impress and losses. For those of us over 70, our grandparents were alive when Chief Charlo's band was forced to leave the valley. North of here, the National Bison Range (now called the Bison Range) was returned to the management of Salish, Kootenai, and Pend d'Oreille people in 2020. That story is explained in the "Flathead Reservation" chapter of this book, but it connects to the Bitterroot in meaningful ways. It shows that we can find redemption for our sins. Injuries, even deep ones, can be softened by shared acts of stewardship.

In 2022, with tribal guidance, the Higgins Avenue Bridge in Missoula was renamed the Beartracks Bridge. This place of expulsion is now honored and partly reclaimed by Salish, Kootenai, and Pend d'Oreille people. A ceremony was held including a parade, drumming, singing, dancing, and emotional speeches. That day, tribal members expressed gratitude for the name change—overdue recognition of a deep injustice. The smiles of elders were real and uplifting. Anger may have been obscured, but new hope was evident. The Beartracks Bridge has been ethically recast. In the Salish language it is called Sxwů̆ytis Smxe.

The story of the Bitterroot Valley exemplifies the larger story of Montana. A Native American world was settled by Whites and freely subdivided until we found a redemptive way forward.

FURTHER READING

"An Agricultural History of the Bitterroot Valley." 2009. Allen Bjergo. www.bitterrootvalley.org.

Bitter Root Land Trust. www.bitterrootlandtrust.org.

Montana Ghost Dance: Essays on Land and Life. 1998. John B. Wright. University of Texas Press.

"The Montana Subdivision and Platting Act in Practice in Nine Montana Counties." 1979. Mark Beardslee. Master's thesis, University of Montana.

Rocky Mountain Divide: Selling and Saving the West. 1993. John B. Wright. University of Texas Press.

Jackleg fence, Big Hole Valley

CHAPTER 7

Land Conservation

The Big Hole Valley is high and remote, surrounded by lofty glaciated mountains—the Pioneers, Pintlers, and Beaverheads. It's a landscape of perfect cattle ranches owned by old Montana families, people who mastered the art of growing irrigated hay in cold, high country. The Beaverslide Hay Stacker was invented here, a wooden frame with pulleys to pile loose hay in stacks 20 feet high. This system is so ingenious that a book was written about the Big Hole titled *Making Hay*. The valley is called "The Land of 10,000 Haystacks," and they remain standing as golden monuments to hard work and wisdom. That's even the name of the valley's largest town: Wisdom, Montana, population 98.

The Big Hole is what most Montana valleys used to be. Productive, wildlife rich, and achingly beautiful with unbroken vistas from river to ridge. Plenty of elk, moose, pronghorn, mule deer, mountain lions, whitetails, black bears, and reports of griz. Cinnamon teals, great blue herons, sandhill cranes, meadowlarks, kestrels, and harriers. Fluvial Arctic grayling live here, the last place in the Lower 48 where this magical opalescent fish is a river dweller; an Ice Age relict in a warming world.

The Big Hole is precious ground. This vast basin saw little development pressure for decades—too cold, too isolated, too many mosquitoes. But by the 1970s, the development boom sweeping the state was poised to come here. Montana's land subdivision law was already proving ill-suited to contend with this land rush.

In response, the Montana legislature passed the Open-Space Land and Voluntary Conservation Easement Act in 1975, many years before most states. It codified land stewardship and brought the Montana law in line with national conservation easement laws. Donations or sales of development rights now brought significant financial incentives. This is a fine example of Montanans' practical nature.

Conservation Easements

A conservation easement is a partial interest in land that perpetually limits subdivision and development on a property with significant agricultural, scenic, ecological, or historical value. Conservation easements are voluntarily donated or sold by a landowner to a qualified nonprofit "land trust" or unit of government. The land remains privately owned and on the tax rolls. Public access is not required. Ranching, farming, and timber management continue as before. The whole point is to keep the landscape as it is.

Landowners receive significant federal and state income tax savings for donating an easement. Federal taxes can be eliminated for up to 16 years and estate taxes can be greatly reduced or wiped out. The "death tax" need not result in the loss of a ranch to the IRS. The value of each donated or purchased easement is different depending on the "before and after" appraisal of the property. What is the value of a property that can be subdivided? What is its value as agricultural land? The difference is the value of development rights that become a conservation easement gift or the value of a sale. Donated easements are treated similarly to money you give to your church or the United Way. In purchased easements, cash received is taxed as a capital gain. In all cases, property taxes continue to be based on the agricultural value of the land.

The concept takes time to fully understand. It seems almost too wise.

In *Saving the Ranch: Conservation Easement Design in the American West*, the authors put it this way: "A conservation easement is a family decision to save your ranch and leave a legacy of good land. With a conservation easement you donate or sell all or most of your development rights to an organization you trust in order to protect specified conservation values. When the conservation easement deed is recorded, those development rights are extinguished. The land trust or agency cannot use them or sell them. They are gone for good. The donation is forever, but you still own the land and can ranch it, sell it, lease it, or give it to your children. The group that holds the easement enforces your wishes, even after your death, and makes sure that no future landowner violates your vision of how the land should be used. The public is not granted access to your land, it remains the landowner's call. The easement is forever and runs with the property. A conservation easement is basically a formal declaration of long-term land stewardship."

Easements arose from a coalescence of private property rights, state law, and federal tax policies. Any one of them by itself was insufficient. It took cooperation to get it done.

In 1977, a forester named Orville Daniels decided to place a conservation easement on his 2,000-acre ranch north of Wisdom. Protecting the place was Orville's legal promise to never develop it. More importantly, his pledge would have to be kept by whomever owned the land in the future. The vow was forever.

In 1978, a group called the Montana Land Reliance (MLR) was established to receive conservation easements like Orville's. MLR is a land trust, a 501(c)(3) nonprofit organization with a mission to help landowners conserve their property forever. They joined The Nature Conservancy in this work but focused on protecting agricultural "working" lands.

At first, opponents saw conservation easements as a government land grab, Stalinism at best. In truth, the tool is based on a deep respect for private property rights. No one can make landowners place an easement on their place.

Despite that, it was slow going at first. Between 1976 and 1988, only 60,000 acres were protected by easements in Montana. In 1989, The Nature Conservancy received a 113,000-acre easement on the Flying D Ranch near Bozeman. Clearly, the benefits of the tool were sinking in. Here was a way to expand a landowner's income without

Ranch protected by a conservation easement

selling the place. Ranchers and farmers could capitalize on their major asset while staying in charge. Today, members of the Montana Stockgrowers Association say they "stand firm on the protection of private property rights" and believe that "voluntary conservation should be encouraged." Across the West, agricultural organizations are active proponents of conservation easements. Members of the Partnership of Rangeland Trusts (PORT) include the Wyoming Stock Growers Land Trust, which has partnered with 90 families to conserve 300,000 acres of working agricultural land, and the Colorado Cattlemen's Agricultural Land Trust, which now holds conservation easements on over 700,000 acres of prime ranch country. Other PORT members are the Rangeland Trust of Kansas, Montana Land Reliance, Nebraska Land Trust, South Dakota Agricultural Land Trust, Northwest Rangeland Trust, and Texas Agricultural Land Trust. The approach is widely seen as damn-right capitalism.

Totaling Things Up

In the past fifty years, 6 million acres of private land in Montana have been permanently protected from subdivision. About 3.5 million of those acres are under conservation easements, with the remaining 2.5 million acres conserved by voluntary purchases by NGOs and agencies. Maps and inspiring stories are revealed in upcoming chapters, focused largely on easements as a creative tool to protect ecologically vital cultural landscapes. We stress land conservation actions by individuals, since they reflect personal and family decisions to live sustainably on the Earth. These choices are compelling and relatively new. Land purchases by NGOs, cities, counties, and agencies have gone on much longer, occurring on high-priority lands where close management is needed. We cover those stories where they have not been explained elsewhere and where they link to major patterns of land conservation. Changes in land status to a wilderness, national recreation area, or national wildlife refuge are based on administrative and political decisions—cases we discuss where they are most relevant in the eight regions we cover.

Montana is a remarkable part of a national story. Conservation easements date back to 1891 and a group called the Trustees of Public Reservations. People in Boston began protecting wetlands, farms, and forests from spreading urbanization. The idea took off. By 1950 there were 53 land trusts nationwide. By 1965, 132 trusts were operating in 26 states. Ten years later, the total climbed to 308. Today, there are more than 1,000 land trusts in the country, 465 of them fully accredited by the Land Trust Alliance.

The Land Trust Alliance reports that 61 million acres have been protected by these creative NGOs as of 2021. Millions more have been saved by city and county open space programs, as well as federal and state agencies. Conservation easements have played a central role across the country. While acreage alone is not an accurate measure of ecological value, it reveals the scale of the achievement. As you'll see, Montana is a national leader in quantity *and* quality.

Three Attempts at Stewardship

Montana has tried three paths to land stewardship that differ in fundamental ways. The state's constitutional right to a healthy environment was groundbreaking, but largely symbolic until a recent court ruling that climate change impacts must be considered in projects reviewed by state agencies. The Montana subdivision law ended up determining *how* land would be developed. The conservation easement law was entirely different. It provided a financially compensating way for landowners to ensure that their land would *never* be developed. The tool was visionary, but businesslike—a way to get things done without fighting about it.

The powerful differences between voluntary land conservation and government land-use planning regulations show us why (table 1).

People concerned with land development now engage in "conservation real estate." This involves negotiating conservation easements where private stewardship is superior; buying and trading land where ecological management is the goal;

TABLE 1. Differences between voluntary conservation and regulatory land-use planning

Voluntary Conservation	**Regulatory Land-Use Planning**
Purpose: conserve land	Purpose: say how land *will be* developed
Respects private property rights	Limits private property rights
Compensating: pays	Noncompensating: does not pay
Permanent: perpetual legal interests	Temporary: laws, regulations, politics change
Negotiated	Mandated
Politically acceptable	Politically controversial
Flexible: fits land and landowner	Rigid: tries to enforce uniform rules
Occurs before decision to develop	Occurs after decision to develop
Rewards independence	Provokes confrontations
Effective at conserving land	Ineffective at conserving land

and mixing and matching techniques in inventive ways, then finding money to make the deals work.

Over the years, conservationists shifted their thinking in essential ways. Some turned away from regulatory planning, others set aside lawsuits, and still more realized that environmental advocacy is best used for public land issues, not private. Land conservation practitioners ended up embracing some very American values:

1. Use the free market. The sacred Earth is also real estate.
2. Honor constitutional rights. Private lands are *very* different from public lands.
3. Respect your neighbors. Voluntary approaches work; regulations don't.
4. Pay for what you want. Financial compensation is required.

In a sense, land conservation became conservative, not always socially or politically, but as an acknowledgment of how the world works. Basically, tools are now matched with the issue at hand. This has a striking correlation to Buddhist teachings about the "Middle Way," pulling back from extremes and embracing life as it is without illusion.

Many NGOs, Agencies, and Tribes

The list of players in this land conservation chronicle is long. Montana has 14 national, regional, and local land trusts. The Nature Conservancy was the first group to operate in the state. Government agencies also played a key role: Montana Fish, Wildlife and Parks, US Fish and Wildlife Service, US Forest Service, Bureau of Land Management, and more. Montana is home to 12 tribes on seven reservations where land conservation efforts are strong. Many of these players will be introduced in upcoming chapters and will reappear where their work moves the story along. For now, we offer a listing of land trusts whose efforts will come to life region by region across the state.

Bitter Root Land Trust: Hamilton
Blackfeet Indian Land Trust: Browning
The Conservation Fund: Missoula
Five Valleys Land Trust: Missoula
Flathead Land Trust: Kalispell
Gallatin Valley Land Trust: Bozeman
Kaniksu Land Trust: Thompson Falls
Montana Land Reliance: Helena

Pronghorn buck

The Nature Conservancy: Helena
Prickly Pear Land Trust: Helena
Rocky Mountain Elk Foundation: Missoula
Trust for Public Land: Bozeman
Vital Ground Foundation: Missoula
The Wilderness Land Trust: Helena

Landscape Ecology

Early conservation easement projects were opportunistic. When someone said yes, a deal was struck. Over the years, easements bunched up—agglomerated—until a strategic vision emerged. This was based on habitats and geography, but mostly on building neighbor-to-neighbor relationships. In time, land trusts became a respected source of useful information, so more landowners decided to place easements on their property. The principles of landscape ecology help guide land conservation work:

1. Large conserved areas are better than small ones.
2. One big conservation area is better than smaller ones of the same total acreage.
3. Wide areas are better than long, skinny ones, reducing the damaging edge effect.
4. Corridors creating connectivity are better than isolated conserved areas.
5. Less-disturbed habitat is better than highly disturbed.
6. Largely intact landscapes are better than fragmented ones.

Woolly mammoths offer a surprising example of the importance of these principles. We usually think this species went extinct long before civilization, with no bearing on conservation today. Not true. Woolly mammoths survived until 4,300 years ago on Wrangel Island in the Chukchi Sea off the coast of Siberia. These animals still roamed while the pyramids of Egypt were being constructed and long after humans shifted to agriculture all over the world. The Ice Age may have ended, but mammoths were still around. So were bristlecone pines in the White Mountains of California. A famous tree was 500 years old when woolly mammoths finally became extinct, and it is still alive. Methuselah is now 4,850 years old based on tree rings. A thousand years isn't so long after all.

But why did mammoths survive so long and then disappear? During the peak of the last continental glaciation, sea level dropped and Wrangel was not an island. It was just another part of Asia where woolly mammoths thrived on expanses of tundra vegetation. But as the ice melted, sea level rose and Wrangel became isolated from mainland hunters. About 300 woolly mammoths became "trapped" on the new island. Over time this small population interbred and their genome lost diversity, accumulating harmful mutations not seen in mainland fossils. Mammoths on Wrangel experienced a "mutational meltdown" caused by a lack of new, adaptive genes. They died off because of lost fertility, finer hair that could no longer keep them warm, a poor sense of smell, and other factors. There were no large predators on Wrangel Island, human or otherwise. Woolly mammoths finally went extinct because the habitat was too small and isolated.

In Montana, the lesson is clear. Conserve large areas and connect them, especially as the climate changes and species need to migrate and adapt. On Wrangel Island, "genetic rescue" was not possible. In Montana, it will be, if conservation projects are designed wisely.

Saving the Big Sky

Visitors to an open Montana valley often remark, "It's so great to see places where development pressure hasn't come yet." Those who know the truth nod their heads and let it go. Many of these largely undeveloped landscapes did not survive by accident. It took decades of hard work by landowners, land trusts, tribes, and agencies.

Saving the Big Sky takes endurance, the Constitution, money, and a deep love of land. The eight regions of Montana we explore reveal inspirational stories. It turns out that people in every setting found a different path to land conservation. But first, we offer an overall sense of the achievement.

What a hopeful tale it is.

FURTHER READING

Conservancy: The Land Trust Movement in America. 2002. Richard Brewer. University Press of New England.

Making Hay. 1986. Verlyn Klinkenborg. Nick Lyons Books.

Montana Ghost Dance: Essays on Land and Life. 1998. John B. Wright. University of Texas Press.

Rocky Mountain Divide: Selling and Saving the West. 1993. John B. Wright. University of Texas Press.

Saving the Ranch: Conservation Easement Design in the American West. 2004. Anthony Anella and John B. Wright. Island Press.

Trust in the Land: New Directions in Tribal Conservation. 2011. Beth Rose Middleton. University of Arizona Press.

CHAPTER 8

Montana: A State of Grace

Such a restorative map.

Large swaths of the state are now conserved in the Rocky Mountain Front, Blackfoot, Greater Yellowstone, Northwest Montana, and other regions. This land will remain ecologically healthy and wisely used.

In the upcoming maps, we use the following legend.

Purple areas are private lands safeguarded by conservation easements—ranches, farms, and forests. Other areas are owned by groups such as The Nature Conservancy, the Trust for Public Land, Montana-based land trusts, and government agencies. Private lands shown in gray remain available for subdivision and housing development.

Dark green areas are formally conserved public lands—federal, state, and locally owned property with a defined conservation purpose. These include national parks, federal and tribal wilderness areas, Areas of Critical Environmental Concern, national wildlife refuges, national monuments, national recreation areas, State of Montana forests and Wildlife Management Areas, and city and county open space lands. It is extremely unlikely that these areas will ever be developed.

Light green areas are public lands serving conservation purposes. These are managed with public input. For example, the Multiple-Use Sustained-Yield Act guides US Forest Service decisions, and the Federal Land Policy and Management Act oversees Bureau of Land Management actions. Some of these properties may be sold, traded, or used in a way that weakens their ecological integrity. These include US national forests and State of Montana forests, state lands, Bureau of Land Management holdings, Wilderness Study Areas, Wild and Scenic River corridors, and designated "Conservation Areas" like the Rocky Mountain Front. However, legislative and political safeguards make widespread conversions unlikely. Adjacent expanses of formally conserved land make wholesale land transformations even less probable. In fact, some light green areas may change management status to dark green in coming years. When land conservation percentages are calculated for each region, these public lands are included.

Light brown areas are shown as tribal lands—Indian reservations with a mix of Indigenous and non-Indigenous ownership. Management is different here than in the rest of the state since tribes have their own land-use systems. Tribal lands are not shown as conserved. However, on the Blackfeet Reservation, the tribe has created a land trust to secure conservation easements on inholdings. These protected lands are shown in purple in the "Rocky Mountain Front" chapter. In other places, formal protections exist, such as the Mission Mountains Tribal Wilderness on the Flathead Reservation.

These definitions and colors are used to focus on landscape-level conservation status. They give a big-picture sense of where we are, while suggesting where further work might be done—where corridors exist or do not, which regions are doing well and which are not. However, this book is *not* a land conservation strategy, and no private or public lands are "targeted" for protection.

We realize that some land ownership details are omitted on this and other maps presented here. This is intentional and serves to communicate the overall pattern by simplifying the cartographic challenge. Showing all ownership details would turn the maps into a cluttered mess. Online sources and atlases can assist people wanting to drill deeper. Our cutoff point for mapping was the end of 2021 because of production logistics. Conveniently, this gave us a roughly 50-year view.

In Montana, 6 million acres of ranchland, farms, wildlife habitats, and open space have been

OPPOSITE: Axolotl Lakes

permanently conserved in the past half century. Again, 3.5 million acres are in easements and the remainder were obtained by purchase or trade for high-priority lands. These properties will never be subdivided and will barely be developed. They will be spared from poor timber management and mining, and forever bonded with water rights. Changes in status on public lands since 1970 are not included in that 6-million-acre figure. However, shifts from light green to dark green (strengthened protection) *are* shown in the pie charts for each region. Those adjustments reinforce our point that Montanans of all political interests are deepening their commitment to land stewardship. Despite that, the need for land conservation increases every year as Montana's population grows.

In 1909, the Montana Department of Publicity saw this coming: "As older states have filled up, the pressure from those seeking homes has become great and the fact is realized that the only area of land in the U.S. waiting to be peopled is Montana. Many large ranches have become too valuable to be used as pasture and have been divided into small tracts and sold." Some things never change.

The land stewardship map of Montana is wise. It shows how much has been done not by opposing growth, but by working hard with the right tools. There is no place in this chronicle for pessimism. There is plenty of room for questions, constructive criticism, and new ideas. Six million acres conserved sounds impressive, but that leaves millions of acres undecided. Therefore, every choice we make shapes Montana forever.

We hope you will be inspired by the following stories about eight remarkable regions of Big Sky Country.

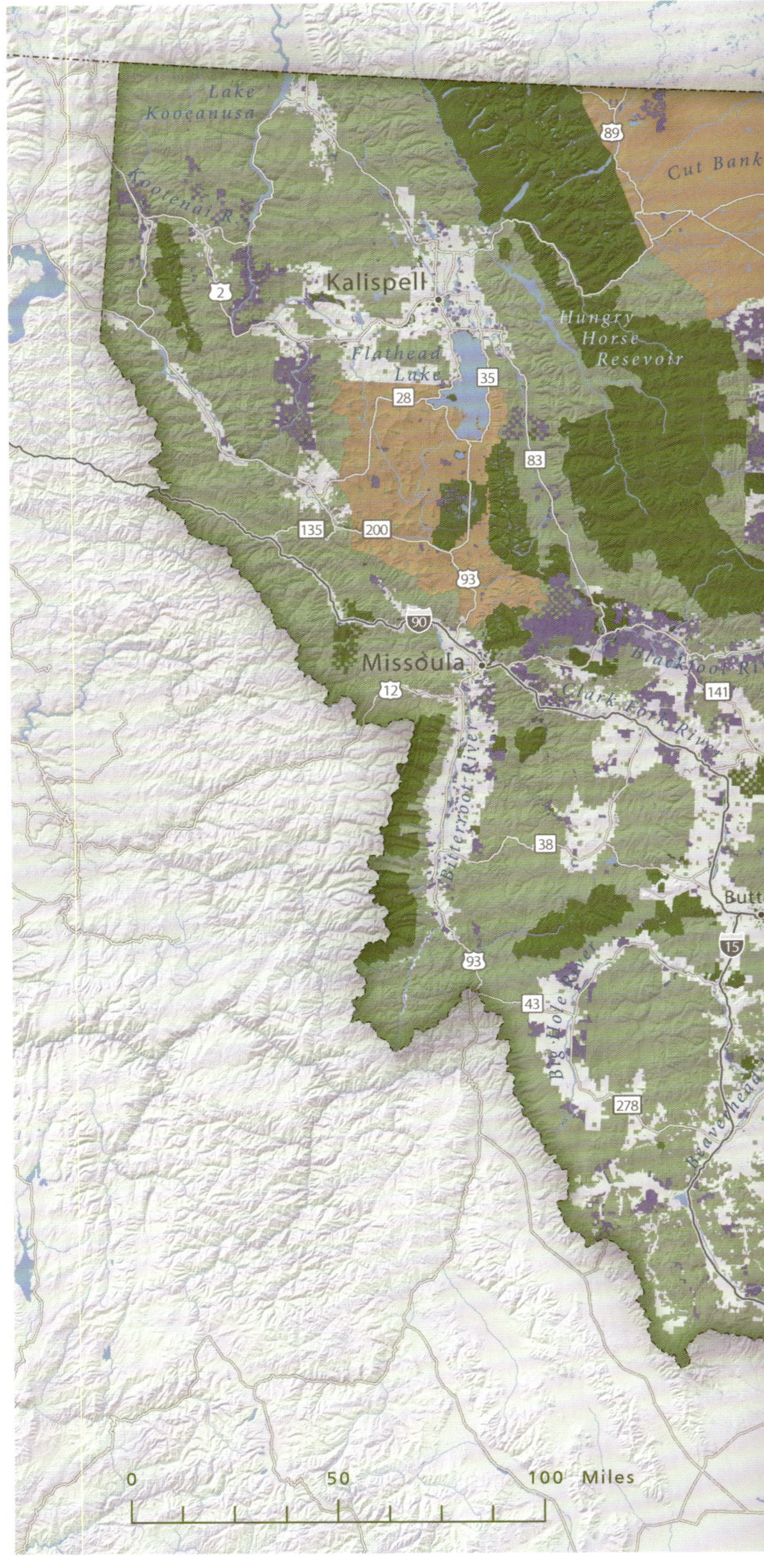

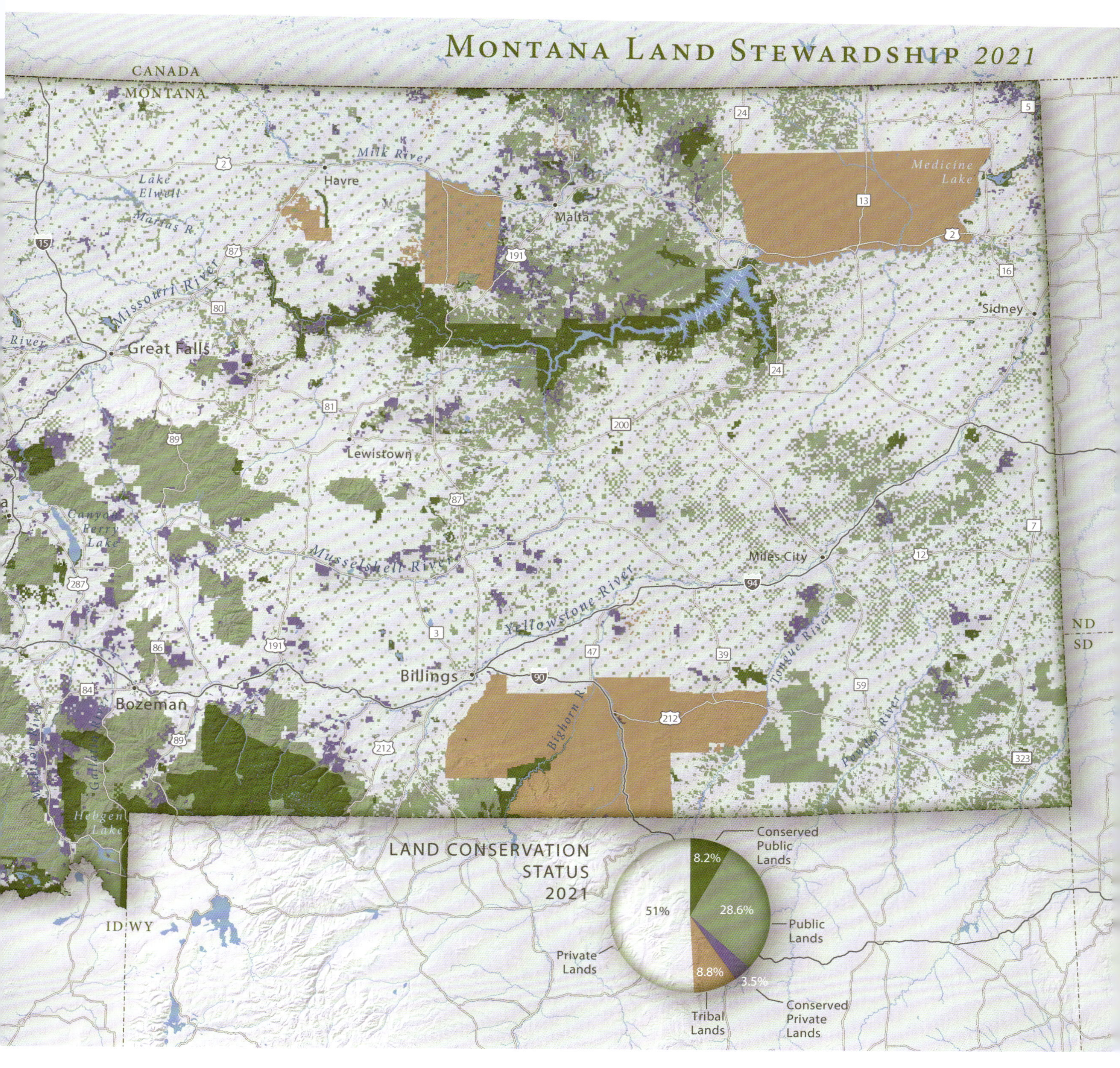

MONTANA LAND STEWARDSHIP 2021
CANADA
MONTANA
Milk River
Havre
Malta
Medicine Lake
Lake Elwell
Marias R.
Missouri River
Fort Peck Lake
Sidney
Great Falls
Lewistown
Canyon Ferry Lake
Musselshell River
Miles City
Yellowstone River
Tongue River
Billings
Bighorn R.
Powder River
Bozeman
Hebgen Lake
ND
SD
ID WY
LAND CONSERVATION STATUS 2021
Conserved Public Lands
8.2%
28.6%
Public Lands
51%
Private Lands
8.8%
3.5%
Tribal Lands
Conserved Private Lands

CHAPTER 9

The Rocky Mountain Front

Wallace Stegner described the place this way: "The Front is at the edge of the mountains but not in them; high plains country, chinook country, its air like a blade or a blowtorch, its sky fitting down close and tight to the horizons and the great bell of heaven alive with light clouds, heat, stars, winds, and incomparable weathers." Those words came from his foreword to A. B. "Bud" Guthrie's masterpiece *The Big Sky*. If you want to learn about the historical flow of people through this landscape, read this novel. At the end, Boone Caudill grieves as Indians get pushed around, animals lose ground, and settlers arrive. "It's all spiled, I reckon. The whole kaboodle." Boone sees his own hand in the destruction of good people and a fine landscape. But with all due respect, while the Rocky Mountain Front region has changed since presettlement times, it is far from spoiled.

Glacier National Park and the Bob Marshall, Scapegoat, and Great Bear Wildernesses form The Front's wild heart. The Blackfeet Reservation is home to hopeful cultural and environmental trends. Well-run ranches spread east from the ridges and roll toward the sunrise, where grizzlies forage on grasslands as they did for millennia before Lewis and Clark feared them. Elk gather in herds so large you stop in your tracks to stare. There are waterbirds, eagles, and songbirds beyond counting. Rivers pour from dozens of canyons extending south from Glacier National Park to Highway 200. They irrigate the plains for livestock and wildlife, nurturing a fine world with precious, clean water.

Thanks go to the foresighted work of The Nature Conservancy, US Fish and Wildlife Service, Montana Fish, Wildlife and Parks (FWP), US Forest Service, and others. The 1970 and 2021 Rocky Mountain Front maps show the striking before-and-after picture of land conservation. Some 46 percent of the landscape is now protected in some way, up from 35 percent conserved in 1970. Some of that lies within the Blackfeet Reservation, but the remainder of land inside the reservation is not considered conserved. Land ownerships are complex, including reservations under the jurisdiction of the federal government, Indian allotments, and land owned by nontribal members.

The story of how and, importantly, *where* 46 percent of the area was conserved is compelling.

A Promising Letter

In 1984, an out-of-the-blue letter arrived at the Montana FWP agency in Helena. It was from the Boone and Crockett Club, self-described as "the nation's oldest wildlife conservation organization." The letter invited FWP to submit a proposal identifying important habitat in Montana that might be purchased as a wildlife reserve to celebrate the club's upcoming 100th anniversary. Boone and Crockett was founded in 1887 by Theodore Roosevelt, George Bird Grinnell, Gifford Pinchot, and other hunters and wildlife enthusiasts. From its beginnings, the club advocated for wildlife conservation, but it had never purchased land to further its mission. Montana's FWP knew its task was to propose something that would outshine anything from other states. It had to stress conserving habitat for big game species known to be priorities for the club, given its advocacy of fair chase principles and its longtime role as the keeper of North American big game hunting records.

Enter serendipity: Montana FWP was aware that The Nature Conservancy (TNC) had recently launched a five-year effort to secure the most significant examples of Montana's biodiversity. This was aimed at celebrating the upcoming 100th anniversary of Montana's statehood in 1989. In these nearly aligned centennial anniversaries of the state and the Boone and Crockett Club, FWP saw an opportunity to propose a habitat conservation

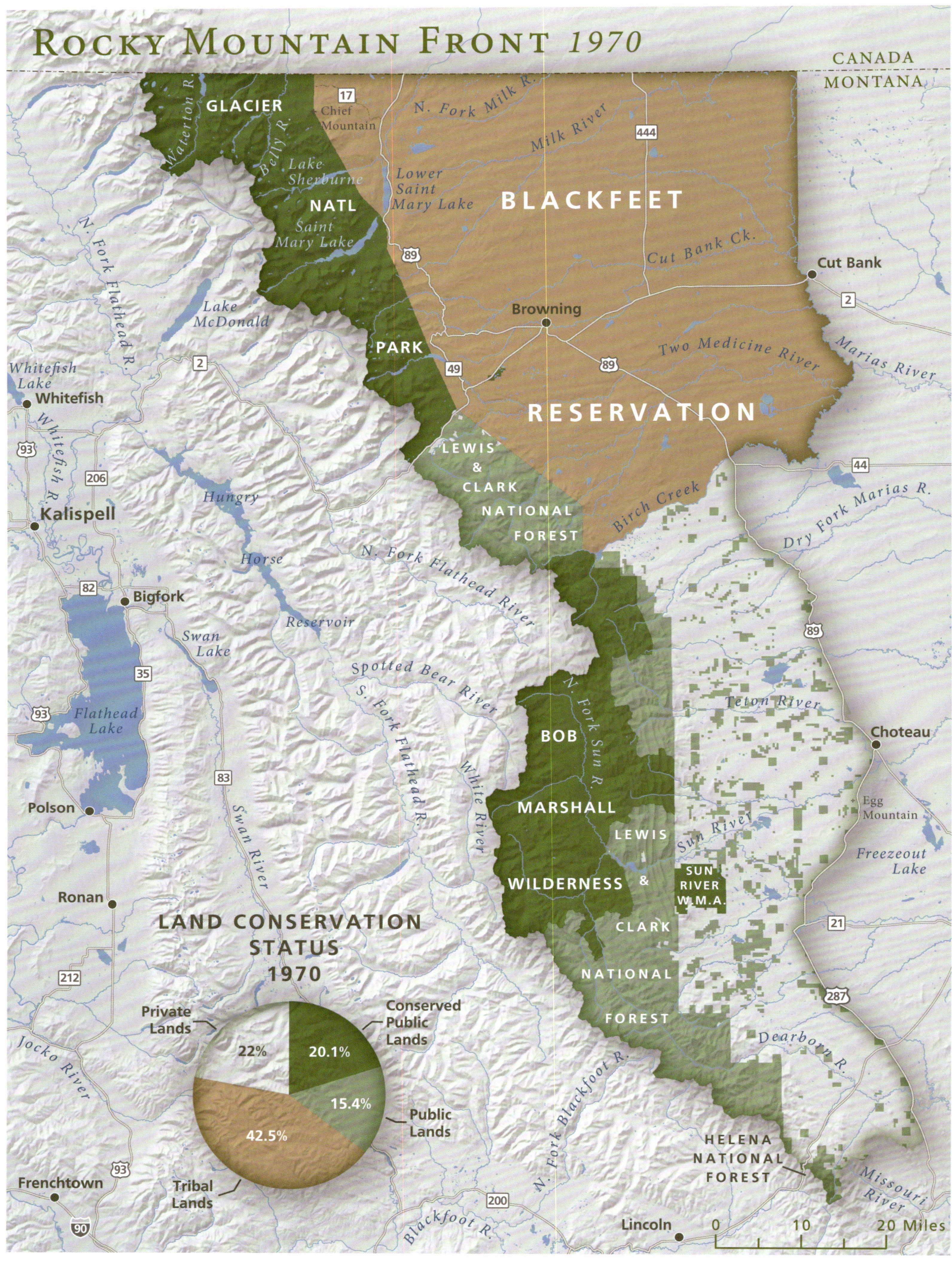
ROCKY MOUNTAIN FRONT 1970
CANADA
MONTANA
GLACIER
NATL
PARK
BLACKFEET
RESERVATION
LEWIS & CLARK NATIONAL FOREST
BOB MARSHALL WILDERNESS
LEWIS & CLARK NATIONAL FOREST
SUN RIVER W.M.A.
HELENA NATIONAL FOREST
Chief Mountain
Lake Sherburne
Saint Mary Lake
Lower Saint Mary Lake
Lake McDonald
Whitefish Lake
Whitefish
Kalispell
Bigfork
Swan Lake
Flathead Lake
Polson
Ronan
Frenchtown
Lincoln
Browning
Cut Bank
Choteau
Egg Mountain
Freezeout Lake
Hungry Horse Reservoir
N. Fork Milk R.
Milk River
Cut Bank Ck.
Two Medicine River
Marias River
Birch Creek
Dry Fork Marias R.
Teton River
Sun River
Dearborn R.
Missouri River
N. Fork Flathead R.
N. Fork Flathead River
Spotted Bear River
S. Fork Flathead R.
White River
N. Fork Sun R.
Swan River
Whitefish R.
Waterton R.
Belly R.
Jocko River
N. Fork Blackfoot R.
Blackfoot R.
LAND CONSERVATION STATUS 1970
Private Lands 22%
Conserved Public Lands 20.1%
Public Lands 15.4%
Tribal Lands 42.5%
0 10 20 Miles

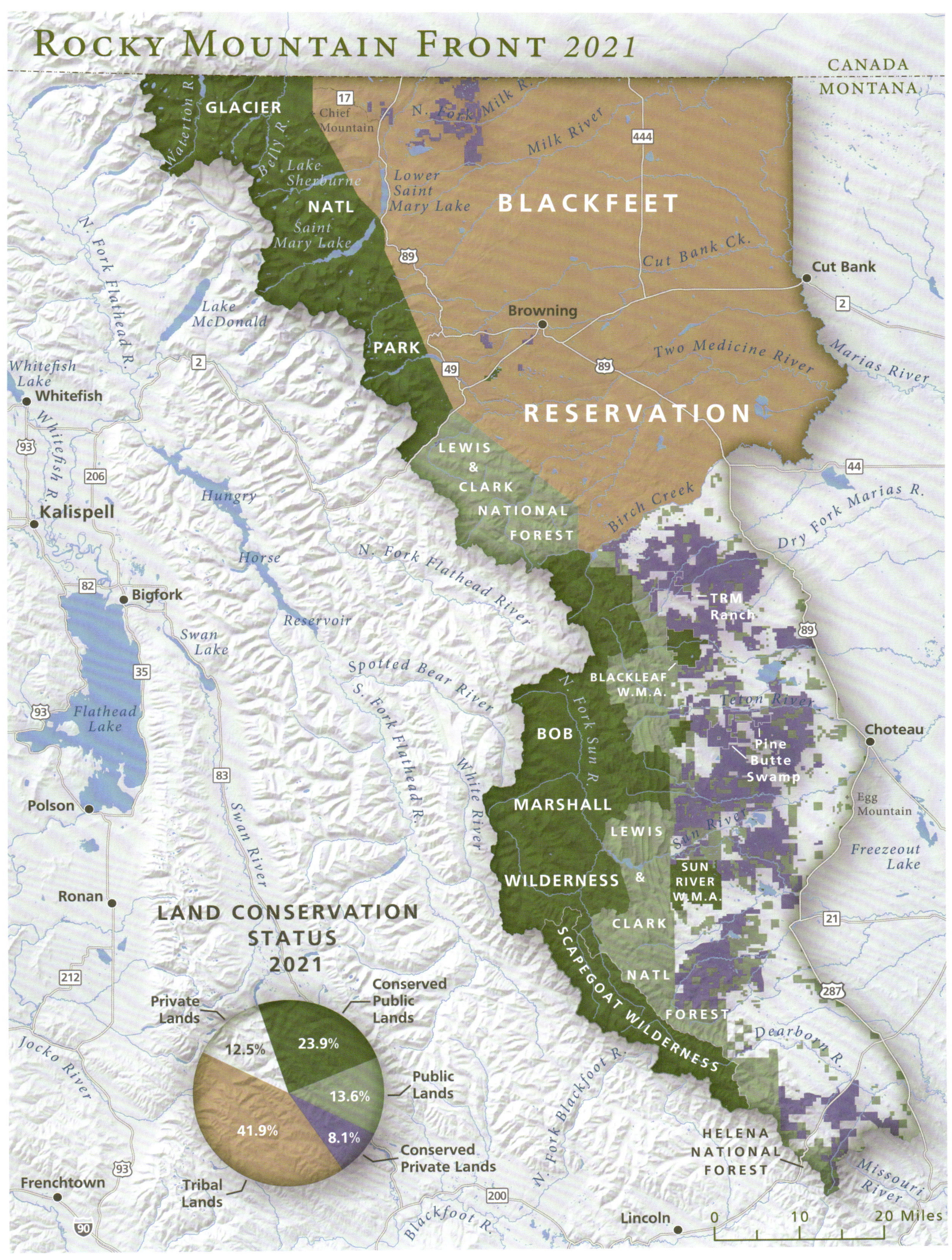
ROCKY MOUNTAIN FRONT 2021
CANADA
MONTANA
GLACIER
NATL
PARK
Waterton R.
Belly R.
Chief Mountain
Lake Sherburne
Saint Mary Lake
Lower Saint Mary Lake
N. Fork Milk R.
Milk River
BLACKFEET
RESERVATION
Cut Bank Ck.
Cut Bank
Browning
Two Medicine River
Marias River
Birch Creek
Dry Fork Marias R.
LEWIS & CLARK NATIONAL FOREST
N. Fork Flathead R.
Lake McDonald
Whitefish Lake
Whitefish
Whitefish R.
Kalispell
Hungry Horse Reservoir
Bigfork
Swan Lake
Flathead Lake
Polson
Ronan
N. Fork Flathead River
Spotted Bear River
S. Fork Flathead R.
White River
Swan River
TRM Ranch
BLACKLEAF W.M.A.
Teton River
Choteau
Pine Butte Swamp
Egg Mountain
Freezeout Lake
N. Fork Sun R.
Sun River
BOB MARSHALL WILDERNESS
LEWIS & CLARK NATL FOREST
SUN RIVER W.M.A.
SCAPEGOAT WILDERNESS
Dearborn R.
HELENA NATIONAL FOREST
Missouri River
N. Fork Blackfoot R.
Blackfoot R.
Lincoln
Jocko River
Frenchtown
0 10 20 Miles
17 444 89 2 49 89 93 206 82 35 93 83 212 93 90 200 44 89 21 287
LAND CONSERVATION STATUS 2021
Private Lands 12.5%
Conserved Public Lands 23.9%
Public Lands 13.6%
Conserved Private Lands 8.1%
Tribal Lands 41.9%

Ram quartet

project that could symbiotically overlap the missions and goals of all three organizations.

The Nature Conservancy was a worthy partner. It was founded nationally in 1951 and has grown to become one of the most effective environmental organizations in the world. TNC currently operates in every US state and 79 countries. Its vision is for "a world where the diversity of life thrives, and people act to conserve nature for its own sake and its ability to fulfill our needs and enrich our lives." Back in the late 1970s, TNC opened an office in Montana with an eye on the landscape where the Rockies meet the Great Plains.

It turned out that the most suitable habitat in the state for a Boone and Crockett Club conservation project existed along the Rocky Mountain Front (RMF, or The Front). The region was rich in habitat for most of the large mammalian species in North America, including grizzly and black bears, elk, and a slowly recovering population of gray wolves, but also for biodiverse wetlands, foothills prairie, willow-lined stream corridors, and cottonwood bottomlands. Where else in the country could you find intact landscapes with such immense wildlife diversity? The only large mammal missing from this landscape at that time was wild bison.

Montana FWP had already established three Wildlife Management Areas (WMAs) along The Front. In 1985, TNC was assembling the Pine Butte Swamp Preserve when it gave a slide show at the annual meeting of the Boone and Crockett Club. More serendipity arrived. After the presentation, one of the club members, Richard Pough, inquired, "Are there native prairie grasslands along the Rocky Mountain Front?" Mr. Pough was the president of the Goodhill Foundation, an entity established to disperse over $50 million from the estate of Katherine Ordway. Ms. Ordway was heir to part of the 3M Corporation fortune. She directed that upon her death, a sizable portion of her assets be liquidated and spent as swiftly as possible on Nature conservation initiatives with emphasis on prairies.

Despite its tough history, The Front had some of the finest prairies left in the West. Learning this, Richard Pough arranged a $1.5 million grant to The Nature Conservancy to work out a collaborative project with the Boone and Crockett Club to secure significant wildlife habitat in The Front. In 1986 the club purchased a 6,500-acre property now known as the Theodore Roosevelt Memorial Ranch as a research, demonstration, and conservation education center. In addition to its wildlife significance, the ranch permanently protects thousands of acres of foothills prairie via a conservation easement held by The Nature Conservancy. Because of its growing engagements in Montana, the Boone and Crockett Club moved its headquarters to Missoula.

Ear Mountain

This "harmonic convergence" of institutions provides a high point where we can see past and present achievements in formal land protection. But no look in the conservation rearview mirror is possible without remembering the sad era of resource exploitation and the assault on Indigenous people and cultures, spurred in part by the notion of "divinely inspired" sea-to-sea expansion known as *Manifest Destiny*.

After the Lewis and Clark Expedition, waves of Euro-American settlers rolled west, and with them came excessive sod busting, fencing, fertilizing, aquifer depletion, overharvesting of forest resources, market hunting, and poisoning of wildlife. Colonization was achieved, but at the cost of considerable decline in the ecological integrity of the region's biophysical landscapes. The collateral damage of genocide and marginalization of Native Americans was the moral price paid. National reckoning with these atrocities has been slow in coming, just as it has with the brutal history of slavery. On the Great Plains astride The Front, vast herds of bison were nearly destroyed, and other populations of fish and wildlife declined precipitously.

By the late 1800s, meat hunters had nearly extirpated much of the remaining wild game herds to help feed the growing population of Great Falls and other settlements. Incursions into the foothills and river drainages to harvest timber for railroad ties and firewood stripped The Front of its low-hanging forest fruit. At the beginning of the twentieth century, only 300 elk remained along the entire RMF. In response, the state legislature established the Fish and Game Department in 1901. By 1941 the legislature had granted the department the authority to acquire and manage important wildlife habitat.

Wildlife Management Areas

In 1948, the Montana Fish and Game Department consolidated parcels to create the Sun River Wildlife Management Area. This embraces nearly 20,000 acres through the purchase and fusion of properties. Originally established to provide crucially important spring and winter range for elk, this WMA is once again home to a diverse array of wild species. Conserving land for an "umbrella species" enabled many other creatures to thrive. In 1976, FWP purchased another 3,000 acres of habitat and named it the Ear Mountain Wildlife Management Area after the iconic mountain peak on its western boundary. This was followed by the Blackleaf WMA, assembled over three years beginning in 1979 and now comprising 10,714 acres.

In 1964, the Land and Water Conservation Fund (LWCF) was established by Congress to fulfill a bipartisan commitment to safeguard natural areas, water resources, and cultural heritage and to provide recreation opportunities to all Americans. The fund strengthens communities, preserves history, and protects the national endowment of Nature. Since its inception, the LWCF has generated over $4 billion for projects in nearly every county in the country.

In the 1970s, the US Fish and Wildlife Service (USFWS) did an inventory of "Unique Wildlife Ecosystems" in Montana. The agency reached out to citizens, clubs, nonprofit organizations, and other agencies, inviting suggestions for areas worthy of permanent protection. Considerable feedback came in touting the Rocky Mountain Front in general, and Pine Butte Swamp in particular, as eminently suitable.

On the strength of this information, in 1978 TNC stitched together private lands (comingled with public lands) into an expanding conservation project. The landscape was a mosaic of prairie, foothills, willow-lined stream corridors, and wetlands with Pine Butte Swamp as the epicenter. Although grizzly bears were the umbrella species,

Sawtooth autumn

securing critical habitat for the great bears allowed the preserve to aid a multitude of other plant and animal species needing the freedom to live, adapt, and evolve. For years following TNC's appearance on The Front, its land acquisition activities were met with suspicion by some landowners and community circles. As the "new kid on the block," TNC was perceived as a national organization with little understanding of neighborhood norms. Critics were incensed that a nonprofit organization could waltz into the territory, buy productive ranchlands, remove them from county tax rolls (not true), and make the neighborhood safer for grizzly bears. For some, the enterprise created conflict with livestock ranching and community traditions.

Egg Mountain

Local criticism softened a little after TNC purchased a fossil-rich site known as Egg Mountain near Choteau. This was the first place in North America where intact nests of dinosaur eggs and the bones of juvenile reptiles were discovered (1979–1980). It revealed colonial nesting behavior and upended scientific understanding of dinosaur nurturing. The species was named *Maiasaura peeblesorum*, the "Good Mother Lizard" found on the Peebles Ranch. The 14 nests and other fossils at Egg Mountain quickly gained national and international fame. In time, the business community in Choteau recognized that world-famous dinosaur digs might be a neighborhood attraction that could "pause" tourists long enough to drop some money in the local economy.

However, runaway rumors and misinformation needed more than dinosaur lore to soften hostility. TNC needed more advocates from the ranching and town communities to support its mission of conserving The Front. TNC's conservation director, Dave Carr, very wisely and capably did so. In 2001 he launched the Rocky Mountain Resource Advisory Council. This group consisted of a local banker (Lyle Hodgekiss) and a networked local attorney (Stoney Burke) along with ranchers, conservation professionals, and agency representatives. They met quarterly to discuss the challenges of creating community cooperation along The Front. The council reached a consensus that the best means of maintaining the livelihoods and lifestyles of ranchers was to conserve the land base by bringing cash into the equation. Community outreach flowing from the council was very helpful in reducing community opposition to formal conservation efforts.

In 1977 a couple of longtime denizens of Choteau, Gene Sentz and Roy Jacobs, a schoolteacher and a taxidermist, respectively, started Friends of The Front. This loose-knit group of citizens was passionate about the wild grandeur of landscape and equally nervous about looming threats to its integrity.

A Strategic Vision

In time, multiple efforts came together as the Coalition to Protect the Rocky Mountain Front, expanding its roster to include more ranchers, business owners, outfitters, hunters, anglers, wildlife lovers, and wilderness advocates. The coalition had a three-group steering committee consisting of the Montana Wildlife Federation, The Wilderness Society, and the Montana Wilderness Association. These groups brought skills to the table in organizing, public relations, and fund-raising as well as considerable savvy about state and national politics. The coalition crafted a strategy that would deal with oil and gas leases, designate more Forest Service land as wilderness, provide more vigorous weed control, and preserve domestic grazing on public lands between the foothills and the ridges. Importantly, it would preserve long-established access to and use of public lands.

The coalition's collaborative efforts produced a piece of federal legislation called the Rocky Mountain Front Heritage Act. It was generated by a diverse mix of local, state, and national citizen constituencies whose efforts were collaborative from the outset. The national forest supervisor, Gloria Flora, made a politically tough decision to remove oil and gas development along The Front. An act was introduced in Congress by Montana senators Baucus and Tester in 2011 and became law in 2014. Once again, the breadth and depth of citizens initiating and promoting this act made it palatable to the politicians whose support was critical. With both wilderness designation and the creation of a Conservation Management Area, 275,272 acres of The Front were protected or on their way.

Snow geese at Freezeout Lake

Private Land Protection

In 2005, the annual field trip of the Montana Wetland Council took place at Pine Butte Swamp. From a prominent overlook, attendees were awestruck by the magnificent surroundings—long views of rolling foothills north and south, eastward vistas over the Great Plains, and towering limestone mountain reefs extending west. This big, intact landscape inspired a big conservation idea.

A joint venture partnership was formed between The Conservation Fund, TNC, the Richard King Mellon Foundation, and the USFWS. The Mellon Foundation placed $15 million in a Rocky Mountain Front account, to be used "creatively" in conserving the most important private lands. Having provided hundreds of millions of dollars for land conservation nationwide, the Mellon Foundation was no stranger to the tools and methods employed by conservation organizations and government agencies It expected a high component of "leverage" in the capital it brought to the table. For every dollar of contributed funds, could the other conservation players produce two dollars? The principal method for achieving this was the purchase of conservation easements on private property, most of them on working ranches.

The partners became a "three-legged stool." Initially, the plan was to raise $30 million in private capital, secure $30 million from the LWCF, and obtain another $30 million from donated conservation easements. TNC and The Conservation Fund took the lead on raising private capital and soliciting easements. The USFWS led the effort to rope in LWCF funds. This cooperation eventually put together nearly $100 million for permanent land conservation. No single partner could have achieved so much alone, but the whole exceeded the sum of its parts, even though each member was different.

The USFWS had to adhere to agency protocols and slower timelines to get congressional approval for LWCF funding. In many cases, ranchers interested in selling easements could not afford to wait for the bureaucracy to play out. In contrast, TNC and The Conservation Fund could move swiftly to provide "bridge funding"—getting easement sellers cash in a timely fashion and later transferring the easements to USFWS when LWCF funding came through. This method of securing deals became a well-oiled machine. The joint venture partners developed a tag-team approach that Gary Sullivan (then with the USFWS, and later with The Conservation Fund) characterized as "the secret sauce on steroids."

Why Conservation Easements?

Giving up some rights of private land use, like subdivision, mining, sod busting, and other uses, seemed to be a reasonable trade-off to landowners who were motivated to protect working ranchlands and preserve a way of life in their blood. Many hoped this culture could continue for their families on home ground they cherished. If a landowner could get paid for a conservation easement, that cash input might well exceed the trade-off of imaginary future value from subdividing, mining, or other exploitive uses. Getting paid for preserving the status quo seemed preferable to placing bets on future economics.

Balancing these trade-offs, a few landowners along The Front signed up to sell conservation easements. That trickle turned into a flood when landowners saw their neighbors benefiting economically from formal commitments to conservation. Donated easements also increased based on expanding federal income and estate tax benefits. Today, more than 300,000 acres of the Rocky Mountain Front are permanently protected by a combination of public and private transactions.

The US Fish and Wildlife Service

This agency was founded in 1940 to build a nationwide system of national wildlife refuges. For decades, the agency drew a hard line around key areas and purchased land within those biologically defined sites. Today there are 567 refuges in the National Wildlife Refuge System, none of which required congressional approval to establish.

But in the 1970s when the USFWS zeroed in on the Rocky Mountain Front, the traditional model of refuge creation did not fit the immense task at hand. The neighborhood consisted of essential spring and winter range for wide-ranging ungulate species, seasonal habitats for large mammals, stream corridors, and wetlands critical for migratory birds. It was improbable that protecting a small national wildlife refuge would really make a difference for all the species that relied on The Front. A new naming convention was needed for USFWS engagement in large landscape conservation. The agency realized it would be far more effective to identify big areas where habitat protection could occur without buying the land. This enlightenment matched evolving conservation science that called for preserving vast landscapes with connecting wildlife corridors.

In 2006, the USFWS named this new territory the Rocky Mountain Front Conservation Area. It was the first such designation in the country, and it gave the agency authority to pursue conservation objectives within a nearly million-acre expanse. One of Montana's senators, Conrad Burns, was chair of the Interior Department Appropriations Committee. Burns made it very clear that he would block any appropriations by the USFWS for land purchases. While he was opposed to more federal land acquisitions, he approved LWCF funds for purchasing conservation easements within the newly delineated conservation area. As a result, easements became the "tool of choice."

The Pace Picks Up

The federal Bureau of Land Management (BLM) then designated ribbons of its land adjacent to national forest lands as Outstanding Natural Areas (ONA) and Areas of Critical Environmental Concern (ACEC). More private lands were acquired or placed under conservation easements by nonprofit organizations. In 1990, the Vital Ground Foundation was born, with the mission "to protect and restore North America's grizzly bear populations for future generations by conserving wildlife habitat and by supporting programs that reduce conflicts between bears and humans." Its first habitat protection project was along the Rocky Mountain Front. More recently, Vital Ground celebrated its 30th anniversary by purchasing a private inholding surrounded by tribal lands east of Glacier National Park and returning it to the ownership of the Blackfeet Nation.

Grizzly Gulch: The Original Creaturefuge

Even kids jumped on the habitat conservation bandwagon. In the late 1990s, the Kratt brothers worked as zoologists producing educational nature shows for PBS and National Geographic. Young viewers continually asked them, "What can I do to help animals?" Chris and Martin Kratt give this account: "Many of our young fans sent small amounts of money from their lemonade stands and allowances in the hopes that they could help protect the wild animals they loved. In 2000, we formed the Kratt Brothers Creature Hero Foundation to empower our young fans. A wide-ranging poll helped us focus our efforts on protecting grizzly bears and other North American creatures. The Creature Hero Foundation partnered with kids, families, the Gap Foundation, Old Navy, and The Nature Conservancy to purchase and protect 1,670 acres of critical wildlife habitat directly southeast of Glacier National Park." The brothers named it Grizzly Gulch, with its exact location best left unrevealed. This "Creaturefuge" (refuge for creatures) is located within a key transition zone for bears between the mountains and the Great Plains. Grizzly Gulch is a place where grizzlies come down onto the prairies to forage. It is also home to elk, moose, wolves, mountain lions, mule deer, beavers, and a host of other creatures. Grizzly Gulch was the first wildlife refuge in North America established through kid action, though it will certainly not be the last.

Crown of the Continent

The Front's wildland virtues are part of the Crown of the Continent. This concept attracted a coalition of conservation partners interested in protecting mountainous wildlands running from Canada through Glacier National Park, then south through wilderness areas along the Rocky Mountain spine. The Crown Managers Partnership (CMP) is "a voluntary group comprised of federal, state, provincial, Tribal, and First Nation land and resource managers and universities in Montana, Alberta, and British Columbia. We recognize that no single agency has the mandate or resources to wholly

address common ecological challenges throughout the region. We therefore work together across borders to tackle shared ecological challenges and concerns." The group's vision is to "ensure a resilient, connected landscape that supports healthy ecosystems and human communities."

After nearly 30 years, the CMP fosters collective, landscape-scale management guided by science and culture. "We are inspired by the understanding that water, fish, and wildlife do not consider borders, and shared resources require shared management. Also, that Americans, Canadians, and sovereign Indigenous nations can work together to conserve this shared landscape for generations to come." From its beginning, the Crown of the Continent idea was spearheaded by Gil Lusk, the superintendent of Glacier National Park. Gil worked tirelessly to create an organizational structure of federal, state, and provincial agencies, tribes, and universities in the United States and Canada. Wilderness areas, national parks, provincial parks, and other wildlands form their geographic core of interest.

The Crown of the Continent landscape is a globally significant climate-change refuge where native plants and animals and ecosystems remain largely intact, and human communities that have coexisted with them for hundreds of years still flourish. The Crown of the Continent is considered by many climate scientists and biologists to be one of the best hopes in North America for preserving wildlands, wildlife, and the myriad cultures and communities in the region. This matches the goals of yet another conservation organization, the Heart of the Rockies Initiative. It believes that "private lands tend to be lower in elevation and higher in productivity, with the best soils and water, and to be near riparian habitats, with the highest biodiversity. These areas are often the 'link' between blocks of habitat, such as national forest, wilderness or roadless areas."

The Rocky Mountain Front has many advocates.

Of Blackfeet and Bison

The 1.5-million-acre Blackfeet Reservation is a landscape of foothills, prairies, aspen groves, and wetlands on the eastern flank of the Rocky Mountains. Chief Mountain is a sacred outlier of the Lewis Range in Glacier National Park. In the 1990s TNC extended its biodiversity and cultural understanding of The Front to include the reservation. Blackfeet tribal member Elouise Cobell joined the board of TNC and envisioned a future in which the Blackfeet Nation's most precious natural features could be conserved. From this vision came the idea of creating a "land trust" combining the best biological research from western science, the best "tools" of law, and the best tribal knowledge about their home ground.

In 1997, Cobell was awarded a "Genius Grant" from the John D. and Catherine T. MacArthur Foundation's Fellows Program. She reportedly used the money to fund a portion of her lawsuit against the US Department of the Interior. For more than a century that agency had failed to account for billions of dollars it was supposed to collect and manage on behalf of more than 500,000 Native Americans. The lawsuit was settled for $3.4 billion. Some of those funds were used to buy back land within reservations.

In 2000, the Blackfeet Indian Land Conservation Trust Corporation was born. This is the first Indian land trust in the United States, and it has acquired private ground along The Front within the reservation and dedicated it to permanent conservation. One property now bears the Indian name given to Elouise Cobell, Yellow Bird Woman. Helen Augare Carlson (Blackfeet) says that the "Yellow Bird Sanctuary serves as a refuge for the human to heal past trauma by acknowledging those necessary relationships that cross boundaries intended to reserve or limit access to all species." It is a place of biological and emotional healing where traditional ecological knowledge is shared.

In the 1990s, the Blackfeet Nation purchased 100 bison from reserves around the country. As excited as the Blackfeet were to bring these totemic animals to the reservation, they were also concerned about the feasibility of restoring free-roaming bison to the plains. Ervin Carlson, the overseer of the small but growing herd, reached out to FWP wildlife biologist Keith Aune, whose familiarity with the Rocky Mountain Front was second to none. In 2008, Carlson and Aune began a conversation about the feasibility of bison restoration. At the time, other biologists had their doubts since global

"Red Dog" and mama

warming was changing prairies and foothills, further drying up already dry landscapes. South of the Blackfeet Reservation, private ownership patterns did not seem conducive to large-scale bison reintroduction. Although a few small private herds were doing well, it seemed unlikely that bison in large numbers would ever roam freely.

Nonetheless, Aune took the bull by the horns. He pitched the Bronx Zoo (home of the Wildlife Conservation Society) to revive the American Bison Society, an organization begun in 1905 whose intention was to save the bison from extinction. Theodore Roosevelt was its honorary president. "Teddy" and other founding members were motivated as much by hunting as by preventing species extinction. In any case, Aune's pitch worked. The American Bison Society was revived, and efforts began to reconnect bison with Blackfeet landscapes and Native peoples.

The Wildlife Conservation Society provided funding for early restoration work. At the same time, the Blackfeet and other members of the Blackfoot Confederacy (four tribes from Canada and the United States) began to think expansively about bison restoration. Other plains tribes were drawn into the conversation, as were the Salish, Kootenai, and Pend d'Oreille from the Flathead Reservation on the west side of the divide. These conversations ultimately produced a historic cross-border Indigenous pact signed by 13 nations from eight reservations. Commonly known as the Buffalo Treaty, this intertribal alliance aims to restore bison to 6.3 million acres in the United States and Canada. Its vision is "to honor, recognize, and revitalize the time immemorial relationship we have with BUFFALO to once again live among us as CREATOR intended by doing everything within our means so WE and BUFFALO will once again live together to nurture each other culturally and spiritually." Tribes that signed the Buffalo Treaty understood that working together was more powerful than any one of them acting alone. Unity created power and capacity to restore critical habitat and bring the bison home. "WE, collectively, invited non-Governmental organizations, Corporations and others of the business and commercial community, to form partnerships with the signatories to bring about the manifestation of the intent of this treaty."

A Collaborative Approach

Since the first Earth Day in April 1970, an astonishing amount of conservation has happened in the Rocky Mountain Front, with more to come. At last, there are accepted tools and money to empower a robust conservation future. What stands out is the role of citizens coming together, first in small numbers, then in larger groupings, forming action-powered organizations that say we should protect one of the most revered landscapes in Montana. The Nature Conservancy established its reputation here, and today state director Amy Croover reports great successes in places like the Crown of the Continent, Northern Plains, and Upper Missouri. TNC now holds conservation easements on over half a million acres, owns 55,000 acres, has led conservation buyer efforts on 113,000 acres, and protected another 650,000 acres in cooperative projects with other groups and agencies. This 1.3-million-acre total is a testimony to TNC's skill and dedication.

However, the protection of the Rocky Mountain Front has been a labor of love for a diverse collection of people. All were willing to put their heads and hearts, talents and voices into figuring out ways to animate land conservation as the dominant value for the neighborhood. Many working ranches and farms have been protected along the way, preserving both landscapes and ways of life that can be passed to coming generations. Humans and wildlife coexist here because of the integrity of Montanans from all backgrounds.

The principal goal of this book is to tally and describe the conservation success stories in the eight regions of Montana. These tales involve a lot of starts and stops, many obstacles and setbacks, and yet through persistence, a huge number of happy outcomes. We celebrate the positives and remain optimistic, but we are mindful of the threats to protecting what Montanans cherish most—personal freedom, clear skies, clean waters, and coexistence with our plant and animal partners.

The conservation of the Rocky Mountain Front over the past 50 years is among the best evidence we have for that optimism.

FURTHER READING

The Big Sky. 2016. A. B. Guthrie Jr. Mariner Books.

Blackfeet Nation Fish and Wildlife. www.blackfeetfishandwildlife.net

Creaturefuge. Kratt Brothers Creature Hero Foundation. www.creaturehero.org

Crown Managers Partnership. www.crownmanagers.org

Crown of the Continent. www.tpl.com (Trust for Public Land website shows various ongoing projects in this ecoregion).

Crown of the Continent Research Learning Center. www.nps.gov (National Park Service information on this conservation effort).

Heart of the Rockies Initiative. www.heart-of-rockies.org (its diverse programs include land trust coordination, public land conservation, rural development issues, and carnivore conflict resolution).

Montana's Rocky Mountain Front. 2016. Rick Graetz and Susie Graetz. Northern Rockies Publishing.

The Nature Conservancy. www.nature.org (website contains details of its projects on the Rocky Mountain Front and elsewhere in Montana).

North Fork meets the Big Blackfoot

CHAPTER 10

The Blackfoot

Eighty-eight percent of the legendary Blackfoot River watershed is conserved.

That earthshaking figure is worth repeating—*88 percent* of all land is protected from development by conservation easements, ownership by national NGOs, or public land designations. Half of this total is permanently secured by easements, wilderness, the Lubrecht Experimental Forest, state parks, Wildlife Management Area status, or management by The Nature Conservancy. The other half is mostly part of the National Forest System where conversion to development is possible, but highly unlikely compared to unprotected private lands. This is an increase from 49 percent to 88 percent conserved in 50 years.

Visitors to the Blackfoot Valley express relief that "development pressure hasn't come here." We nod our heads and smile, because the truth is more noble than that. It took decades of hard work by hundreds of people to make this seeming miracle happen.

That 88 percent figure comes to life as we drive through cattle ranches full of productive beauty and then stop for a burger at Trixi's Antler Saloon. Or maybe when we float from Whitaker Bridge to Johnsrud Park, navigating rocks and rapids. The realization might arrive wading into a pool like Muchmore, fly rod in hand, feeling the enduring strength of the river. In *Home Waters*, John Maclean writes of the wonder of "being attached to a living but unseen power that lies below the surface." Trout tell us how we treat a watershed. Their health is a virtue, their decline a vice.

Much is made of the 30×30 goal of safeguarding 30 percent of the planet by the year 2030. In the Blackfoot, that figure is in the rearview mirror, and everything has happened voluntarily or through a public land management process. There was no regulatory scheme and no confiscation of private property rights. This natural and cultural endowment was honored by listening to each other, showing patience, and finding a respectful way forward.

But why did it happen *here*?

The Blackfoot River Valley is not near a national park. The mountains of the Scapegoat Wilderness are beautiful but lack the scenic drama of the Mission Mountains, Rocky Mountain Front, or Madison Range. There is no Greater Yellowstone concept or American Prairie initiative. No single group does all the conserving. No signature photograph instantly identifies the place, as it does at Many Glacier or the Sleeping Giant. Lower reaches of the Blackfoot are narrow and rocky, unlike other Montana rivers, which meander across wide sunny valleys. You would think that the film *A River Runs through It* would have created a unifying sense of the Blackfoot, but every frame was a mirage. That movie was shot on five other Montana rivers, because in 1991 the drainage was too logged, mined, and trout impaired to play itself.

So, why is *this* watershed so extensively protected?

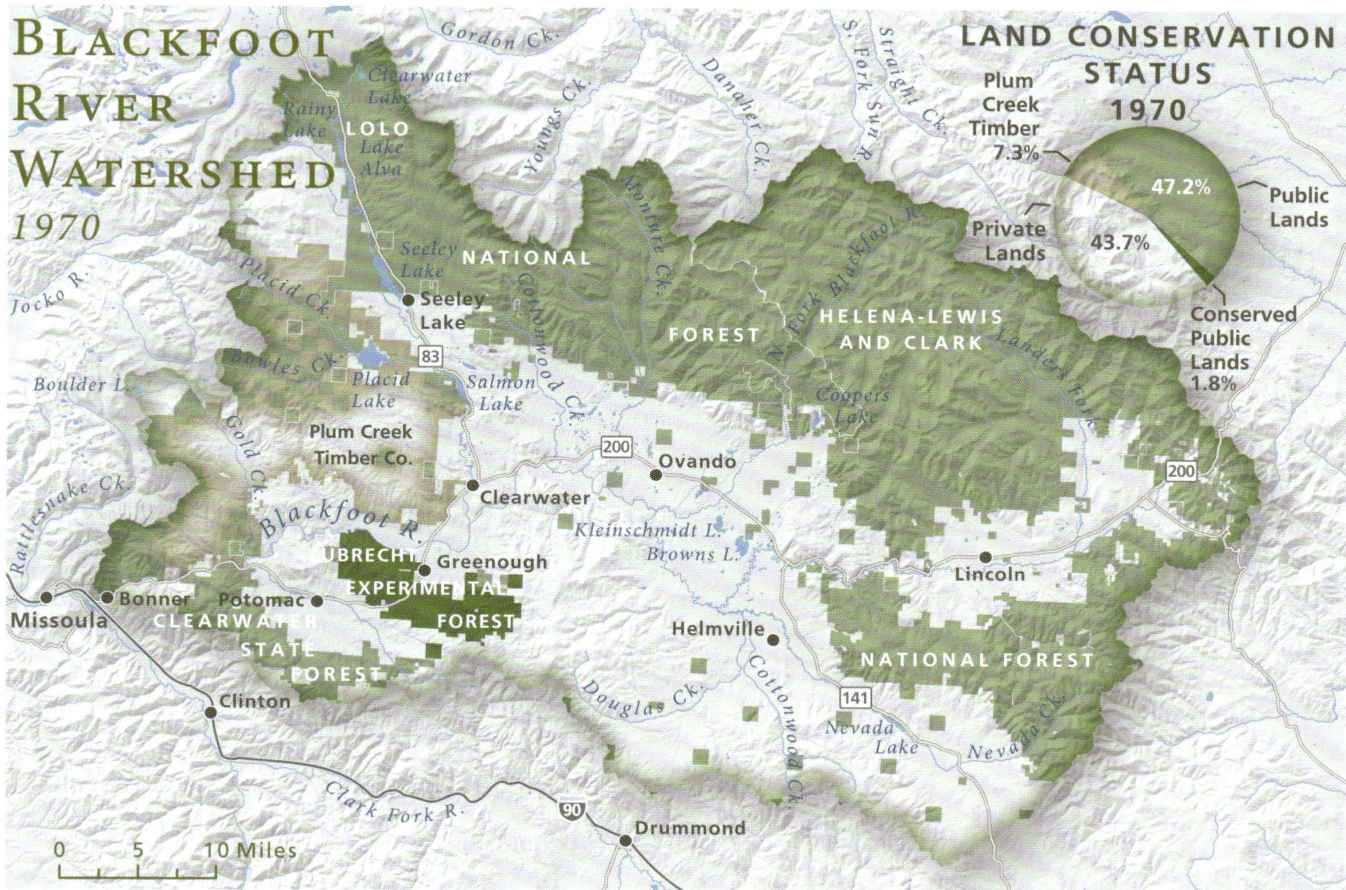

Environmental History

The Blackfoot is a landscape born of profound ecological, cultural, and personal kinships. That's true across Montana, but an acted-on sense of community defines this place. The Blackfoot is about connections and cooperation, actions that build trust. Something rare in any era, but especially now.

Scenically dramatic places announce themselves at majestic volumes. We feel awed and overwhelmed by the spectacle. The Blackfoot speaks in a softer voice, a friend we rely on. We hear its reassurance in the quiet spin of a back eddy, the whoosh of raven wings, and the bawl of a calf. Small gifts offered that elevate our souls. There is room to make peace with ourselves and become our better angels. People don't come to the Blackfoot for the view; they come to experience a worthwhile life. A fortunate few call it home.

This durable geography is built of billion-year-old rocks from the bottom of a freshwater sea. Mud cracks and ripple marks remember vanished watersheds—maroon, green, and gold messengers from a time before dinosaurs and birds. Glacial Lake Missoula filled the basin up to Ninemile Prairie until its ice dam broke and the torrent excavated a canyon now beloved by fishers and floaters. Forested mountains rise up from the water announcing the seasons. Larch trees turn light green in the spring, deep green in the summer, and gold in the fall before sharing an evolutionary secret and dropping their needles. Larch are *deciduous* conifers, something magical. Above Roundup Bridge, the valley is wider, grassier—proven ranch country. Some of those spreads were built by families whose legacies spur land conservation today. The Blackfoot River itself is born in the wilderness, irrigates fields, and then grants us recreation to restore our spirits.

Norman Maclean distilled all this perfectly: "Eventually all things merge into one, and a river runs through it." That novella is a lyric poem, but Norman was hardly the first to bond with the landscape.

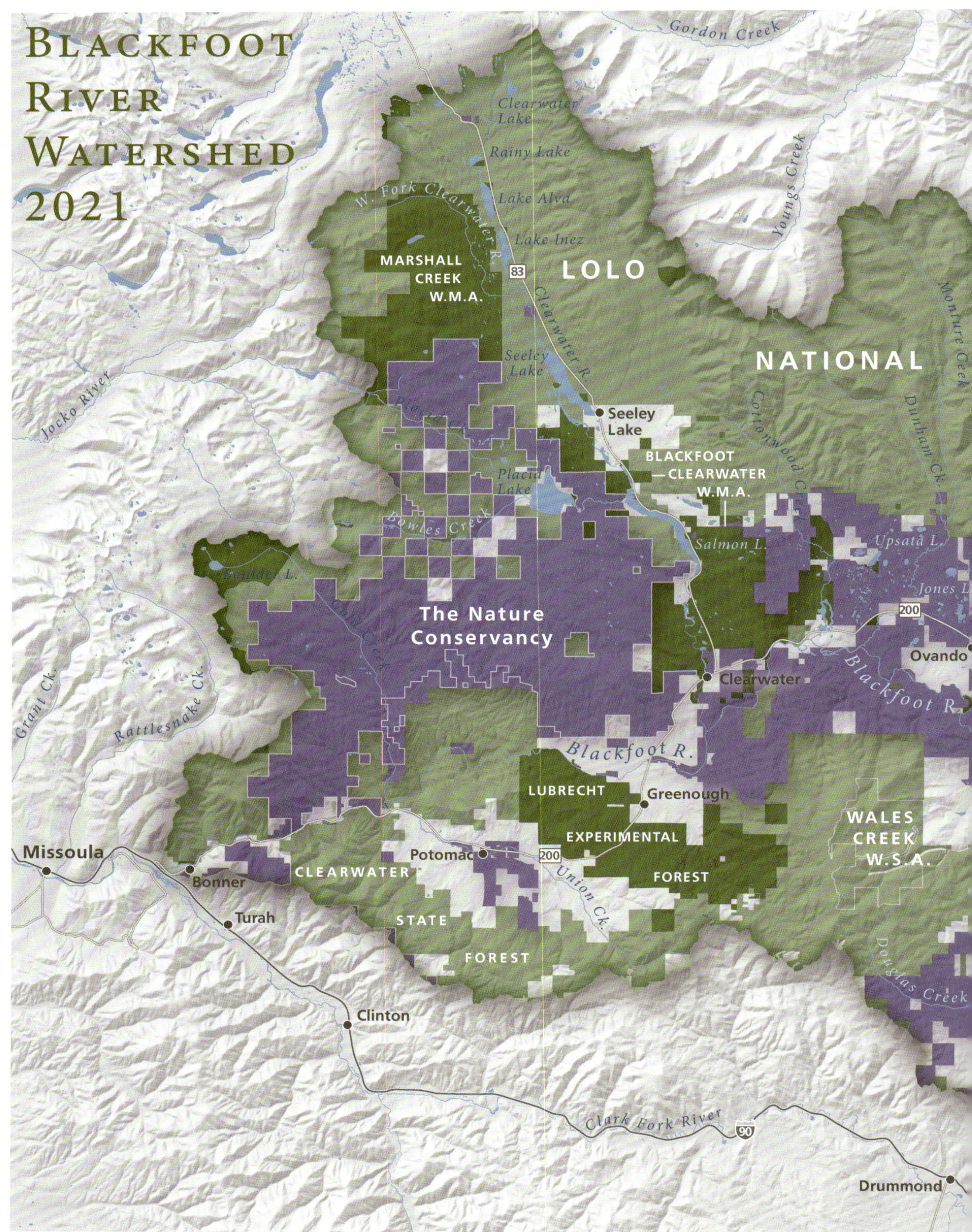

Blackfoot River Watershed 2021
Gordon Creek
Clearwater Lake
Rainy Lake
Lake Alva
Lake Inez
W. Fork Clearwater R.
Youngs Creek
MARSHALL CREEK W.M.A.
83
LOLO
NATIONAL
Clearwater R.
Seeley Lake
Monture Creek
Jocko River
Placid Ck.
Seeley Lake
Cottonwood Creek
Dunham Ck.
BLACKFOOT CLEARWATER W.M.A.
Placid Lake
Bowles Creek
Salmon L.
Upsata L.
Boulder L.
Jones L.
200
The Nature Conservancy
Cold Creek
Ovando
Clearwater
Blackfoot R.
Grant Ck.
Rattlesnake Ck.
Blackfoot R.
LUBRECHT
Greenough
WALES CREEK W.S.A.
EXPERIMENTAL
FOREST
Missoula
Potomac
200
Bonner
CLEARWATER
Union Ck.
Turah
STATE
FOREST
Douglas Creek
Clinton
Clark Fork River
90
Drummond

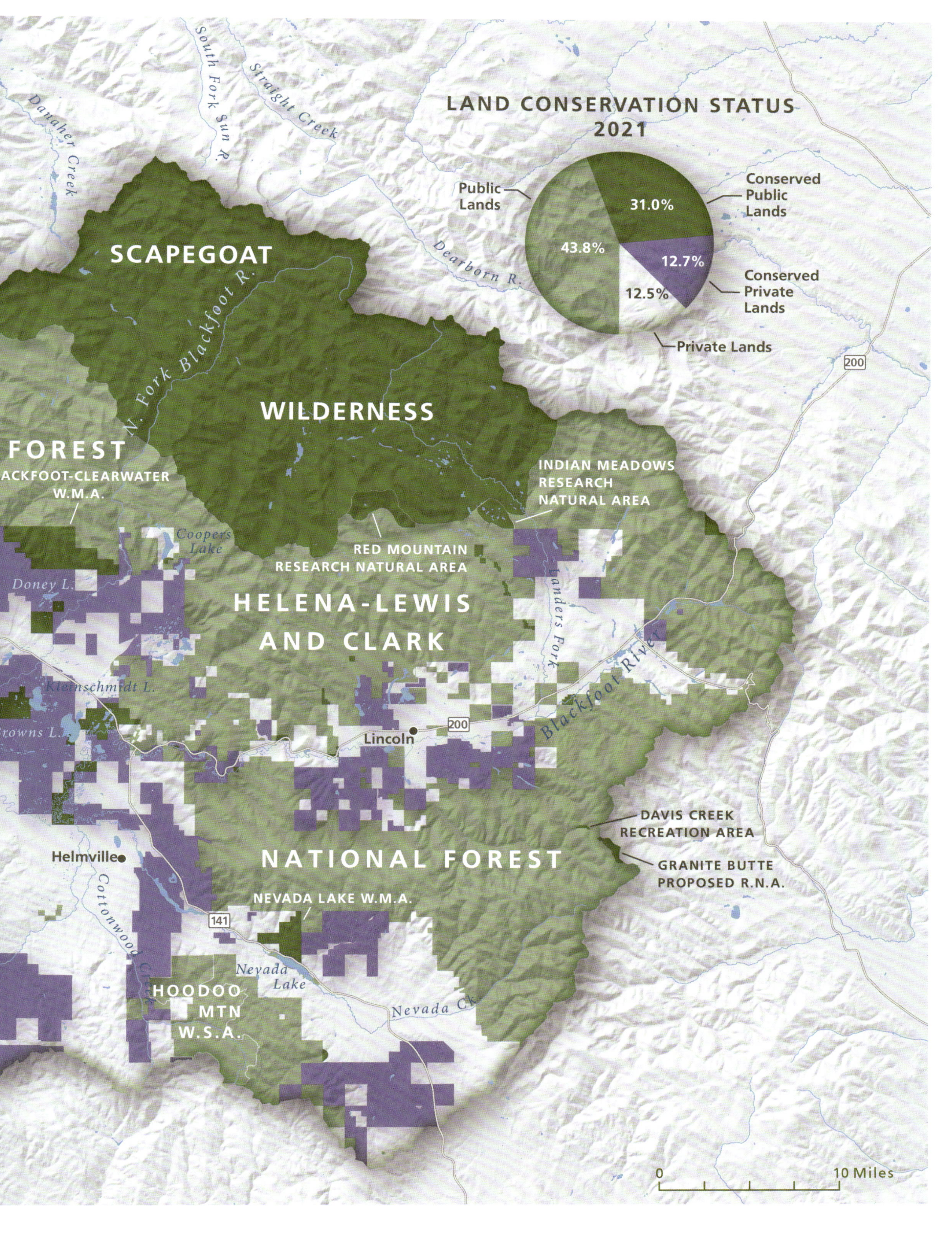

LAND CONSERVATION STATUS
2021
Public Lands
43.8%
Conserved Public Lands
31.0%
Conserved Private Lands
12.7%
Private Lands
12.5%
SCAPEGOAT
WILDERNESS
FOREST
BLACKFOOT-CLEARWATER W.M.A.
INDIAN MEADOWS RESEARCH NATURAL AREA
RED MOUNTAIN RESEARCH NATURAL AREA
HELENA-LEWIS AND CLARK
NATIONAL FOREST
DAVIS CREEK RECREATION AREA
GRANITE BUTTE PROPOSED R.N.A.
NEVADA LAKE W.M.A.
HOODOO MTN W.S.A.
Danaher Creek
South Fork Sun R.
Straight Creek
Dearborn R.
N. Fork Blackfoot R.
Coopers Lake
Doney L.
Landers Fork
Blackfoot River
Kleinschmidt L.
Browns L.
Lincoln
200
Helmville
Cottonwood Creek
141
Nevada Lake
Nevada Ck
0
10 Miles

The Salish and Pend d'Oreille Nations call the Blackfoot Cokahlarishkit—"The River of the Road to the Buffalo." The valley was a vital meat and medicine trail connecting the Northern Rockies to the Great Plains. Hunts were essential since bison were uncommon west of the divide. Travois carrying meat were dragged by dogs, and after 1700 by horses. You can still see those marks inscribed into the prairies. Tepee rings, projectile points, awls, beads, and other Native signatures are found on riverside lands. In *Home Waters*, John Maclean (son of Norman) recounts a trip he made following Cokahlarishkit with Ron Cox, a forester and historian. Remnants of the "road" still existed, as did marker stones, and a storied aspen grove in Ninemile Prairie. Clones like this took root thousands of years ago, reproduced by sprouting, and became essentially immortal. When Meriwether Lewis passed through the valley in 1806, his party had no trouble finding the entire 132-mile route based on these landmarks.

The River of the Road to the Buffalo was also traversed by the Kootenai and Nez Perce. Curiously, this waterway was never a core of the Blackfeet homeland. In 1831, the river was named the Blackfoot by an Irishman named John Work, a major trader for the Hudson's Bay Company. His misleading name stuck on western maps. Fur traders like Work told stories about fights between Rocky Mountain and Great Plains tribes, especially in Hellgate Canyon near Missoula, but the conflict was broader than that. Between 1845 and 1874, Great Plains tribes raided Salish and Pend d'Oreille settlements across Western Montana. War took such a heavy toll that the Salish sought Christian missionaries, who formed St. Mary's Mission in the Bitterroot Valley and the St. Ignatius Mission in the current Flathead Reservation. Tribes sought both additional spiritual power and a practical way to deal with colonialism. The Blackfoot Valley itself was never densely settled by Natives—it was a landscape that connected people and places for good and otherwise.

Copper mining in Butte first linked the Blackfoot to the outside world. In 1884, the Anaconda Company began clear-cutting forests in the watershed mostly to fuel its massive Washoe smelter. A subsidiary, the Montana Improvement Company, set up shop in Bonner. The mountainsides were stripped of trees by crews using crosscut saws to fell immense ponderosa pines, Douglas firs, and western larches. Log drives choked the river until railroads and truck hauling took over. Ecosystems were seen as board feet of timber, and sediments washed down from denuded mountainsides, filling the river with mud. However, we cannot dismiss the economic and social benefits of resource use. Logging and mining created jobs. Lumber built homes, stores, and factories. Copper electrified and plumbed America. Copper was a vital component of shells and other ordnance that won two world wars.

Near Lincoln, the Heddleston Mining District produced lead, zinc, silver, gold, copper, and molybdenum. The Mike Horse Creek watershed was heavily mined until the 1950s, but the environmental bill came due in 1975 when a flood washed out a retaining dam. Three hundred thousand tons of toxic tailings were swept into the Blackfoot River, killing rainbow, cutthroat, brown, and bull trout for miles downstream. In the 1990s, Phelps Dodge proposed excavating a gigantic open-pit gold mine in the headwaters. Poisonous cyanide mist would be sprayed on low-grade ore, dissolving rock and leaving gold behind. Phelps Dodge claimed there would be no impact. This obvious threat to the Blackfoot unified ag people, anglers, and city residents across the state until a state law was passed banning this mining technique. Montanans had seen enough damage to clean water. Most of the Clark Fork watershed was a Superfund site in the aftermath of copper mining operations in Butte and Anaconda. Over many years, the Clark Fork Coalition in Missoula succeeded in ensuring that much of the damage was repaired.

In 2002, The Nature Conservancy bought 117,000 acres of former Anaconda Company timberlands with the goal of moving them into public ownership with the US Forest Service, Bureau of Land Management, and Montana Department of Natural Resources and Conservation. This allows forests to mend and gives species corridors for responding to climate change. But the origin story of land-saving in the Blackfoot was homegrown and started decades before.

Beginnings

In the 1960s Myrna Loy, Charles Lindbergh, and Arnold Bolle met in met in New York City to talk about options for dealing with land subdivision in the Blackfoot—an actor from Helena, an aviator from Connecticut, and a forester from the University of Montana. "Arnie" Bolle proposed buying 1,500 acres of subdivided land inside the Lubrecht Experimental Forest, a 21,000-acre treasure. Money was raised and the first land conservation deal was completed in the watershed. A modest beginning to an epic tale.

The Blackfoot is a community of loners sharing a deep fidelity to the Earth. Hank Goetz and Land Lindbergh are among them. For decades, Hank supervised the Lubrecht Experimental Forest, and Land ran the Lindbergh Cattle Company. Both sought solace in the peace and quiet of the Blackfoot, but they feared that a land development boom was coming to the drainage. Hank is a Vietnam combat veteran, and Land sought a meaningful life away from the shadow of a complicated father. They forged a 60-year friendship based on a shared devotion to the landscape. In the 1970s, meetings were held at the Sunset Hill School that attracted neighbors from all over the Blackfoot. Land Lindbergh wanted a place for ranchers, loggers, outfitters, recreationists, and small business owners to respectfully talk about the future of the place they shared. She said, "The river will tell us how we're doing." A rancher named Bill Potter was also a major force. He ran the E Bar L Guest Ranch nearby. Jim Masar, a ranch hand for Potter, went on to get a law degree and got deeply involved.

Everyone agreed that land subdivision and development was a major threat to the watershed. However, they didn't want zoning or regulatory land-use planning to interfere with their freedom. They also didn't want the Blackfoot designated a Wild and Scenic River since locals mistrusted regulations handed down from Washington. They decided to save the watershed themselves.

A baseline of public lands already existed. In 1937, the Northern Pacific Railway donated land to the University of Montana to create the Lubrecht Experimental Forest as an outdoor ecology laboratory. The 44,000-acre Blackfoot-Clearwater Wildlife Management Area was formed to provide elk winter range comanaged with cattle grazing. The 1.5-million-acre Bob Marshall Wilderness was designated in 1940 to honor a Jewish forester from New York City who hiked endlessly through the mountains. In 1964, it was formally added to the wilderness system. "The Bob" has long been a big part of Montana's wild identity. In 1972, the Scapegoat Wilderness was established. State parks were formed at Salmon and Placid Lakes. In time, the Seeley Lake Game Preserve and the Nevada Lake Wildlife Management Area were created. A wealth of public lands linked peaks to creeks, forests to wetlands, winter range to ranches.

In many places, that would be enough, but in the Blackfoot, things were just getting started.

Darrell Sall of the Bureau of Land Management saw the need to better manage recreation on federal lands along the river. Darrell believed that collaborative ecosystem management was the future—a bold stance at the time. Ernie Corrick of Champion International (a timber company) stepped up because he saw the crucial role its 680,000 acres played. Champion provided logs to Bonner, chips to the pulp mill in Frenchtown, and mine timbers to operations around the state. Ernie felt that clear-cutting should evolve into forest and land management to keep things viable, so he quietly set aside a stand of old-growth timber up Gold Creek. A state forester named Steve Wallace believed that the Clearwater and other state forests could also be better managed. Their involvement, plus the leadership of Hank Goetz at Lubrecht, helped heal divisions between conservationists, forest managers, and ranchers.

Jerry Stokes of the Bureau of Outdoor Recreation then created a plan for recreational use of private land. The US Fish and Wildlife Service (based on successes in the Rocky Mountain Front) then designated the Blackfoot eligible for cash from the Land and Water Conservation Fund to buy easements on ranches. Millions were raised from receipts of offshore oil and gas development and redirected for conservation. Across the watershed, private citizens and agency personnel brought business, administrative, political, and mapping skills forward. Montana Fish, Wildlife

and Parks saw this rare level of engagement as an opportunity for conserving habitat while providing public recreation.

All this made Hank Goetz optimistic. "I learned that my neighbors had the same values I did, but I didn't know it until we started working together. The beauty of the place bonded us." Land Lindbergh agreed but felt that "we needed some guidance because we all had these great ideas."

Ken Margolis and Huey Johnson of The Nature Conservancy were invited over to explain how conservation easements work. This voluntary, financially compensating tool ended up matching the homegrown approach local people favored. Susie Lindbergh was a strong supporter of easements, based on her ethics and education in environmental issues. Soon, a river corridor was collaboratively mapped for protection. The idea was to negotiate agreements with all willing landowners—including the State of Montana—and then put all easements into effect at the same time. People figured they would commit when everyone else did. A grand simultaneous closing would have everyone's back.

It didn't happen. Landowners had different economic needs and readiness to act, so easements came in one at a time. The Nature Conservancy and the Montana Land Reliance built relationships with people, gained trust, and closed deals with those who said yes.

Conserving the Blackfoot watershed began opportunistically.

Early Conservation Easements

Edna and Paul Brunner were pioneers in 1976. They acquired a ranch from a Watergate conspirator and chose to keep it safe forever. Both were direct people, sure about their decision to donate an easement to The Nature Conservancy. Today, the Cliff Ranch is in different ownership, but the Brunners' vision of land stewardship remains. A scarp stares down on a well-tended landscape of grasslands, forests, and glacial pothole lakes.

Joe McDowell was crucial in advocating for easements here and across the state. Born in Montana in 1912, Joe worked as a Justice Department lawyer in DC until at 58 he "came home" to live beside the North Fork of the Blackfoot River. He used his experience with "scenic easements" in the Great Smoky Mountains to successfully advocate for Montana's 1975 conservation easement law. Hank Goetz calls Joe "one of the grandfathers behind that legislation." Land Lindbergh says, "He was really the leader, particularly in the politics. And he was famous for his buffalo steaks and old fashioneds."

Fanny Steele owned the 320-acre 5-Star Double R Ranch on Arrastra Creek south of the Scapegoat Wilderness. Because of an easement and good management, the property remains an ecological treasure. Moose and elk are plentiful, nesting ospreys are loyal to sheltering trees, and polite grizzlies visit in the spring. The sky is full of birds: kestrels, sharp-shinned hawks, harriers, long-eared owls, and sieges of great blue herons. Fanny Steele loved this place of sanctuary. She was a champion bucking horse rider, appeared in movies beside Rudy Vallee and Montie Montana, and used her outfit as a guest ranch with a shed full of tack, rodeo flyers, and movie posters. Fanny was famous for standing up on the haunches of a galloping horse and looking backward while twirling lassos with both hands. A family named Rosenthal ended up donating the easement when Fanny passed, but her presence moved the deal along.

Cora Barbour shared that sentiment. She donated a conservation easement on the 3,656-acre Monture Hereford Ranch. Meriwether Lewis named Monture Creek "Seaman's Creek" after his beloved Newfoundland dog, but that name never stuck. The ranch contains a rare plant called Howell's gumweed (*Grindelia howellii*). Common names are tricky—a rare weed? Cyndi McAllister of The Nature Conservancy found a colony of these yellow-flowered plants being happily pollinated by fat bumblebees. Those meadows are also nesting ground for sandhill cranes and bobolinks. Pothole lakes welcome four kinds of teals. Long-billed curlews search the grasslands for bugs, moving stalks aside with long, down-turned beaks. Biologist Rollie Redmond filled us in on what habitat the curlews need.

Jenny Eder (Salish) and her husband, Otto, heard the news and placed an easement on their beloved 1,280-acre ranch. In *Montana Ghost Dance*, the author describes the project. "The Eders' equipment was old and their yields average. Then developers began circling, carrying hefty checks. The Eders just smiled and gave up a fortune. The small house was filled with the kindest silence I've ever heard. Their pack of ranch dogs pried the screen door open with sharp, intelligent muzzles then walked inside wagging their entire bodies with the certainty of well-loved children."

Marguerite Heller was another conservation pioneer in the Blackfoot. She owned 500 acres, acquired back when land was dirt cheap. Ms. Heller often mused about the philosophical riddle of land tenure—a term derived from the French verb *tenir*, meaning "to hold." She saw ownership as land stewardship, a way to give back to the Earth. The Heller place was a mix of cottonwood floodplain forests and fertile valley bottoms. She lived in a modest log home for many years following a career teaching school in faraway Chinook, Montana. Marguerite was in her 80s (in the early 1980s), living off a small pension and a smaller Social Security check. Her fidelity to the land made up for all that. She had a strong, unflagging love affair with her piece of paradise and spent days immersed in Nature. Marguerite was a keen observer of wildlife, noting their seasonal rhythms and matching them with her own.

Margurite Heller knew that the extraordinary beauty and gentle topography of her land made it an attractive target for subdivision. Over the decades she had been wooed by enough buyers that she could "sense a subdivider" as soon as one rattled her cattle gate. Ms. Heller had family members who stood to inherit the land. She knew they were only interested in maximizing the economic value of their eventual inheritance. The idea of a conservation easement perfectly fit her values and concerns. She connected with The Nature Conservancy and an easement was crafted.

But an obstacle emerged. The Montana conservation easement law stipulated that county planning officials be informed of every deal. The Heller place was in Powell County, and its planning board knew nothing about this new technique. The board members vigorously objected to the very idea that a landowner could restrict her property in perpetuity. To them, it didn't seem right that someone could "tie the hands of future generations." The board never saw the irony in its routine

Graceful trumpeter swan

approval of subdivisions that committed a ranch to development forever. A few people in Montana still don't get it.

In time, reason prevailed, the conservation easement was recorded, and Marguerite's wishes for the land were carried out. She wasn't able to take advantage of the tax incentives for placing the land under easement protection. A retired teacher didn't have enough income to "shelter." Her easement donation came straight from the heart, an act of philanthropy in its purest form. Sure enough, when Marguerite Heller passed away, her heirs sold the place. That was their right, but the conservation easement agreement remains in place. From single acts of grace like this, a broad pattern of land-saving emerged up and down the Blackfoot River. Ms. Heller would be pleased.

The early days of easements in the Blackfoot had an innocence to them. Tax deductions for easements were very small compared to those of today, so people sacrificed real estate value for soul-deep rewards. Back then, Land Lindbergh put it plainly: "We were trying to adjust to the changes that were coming. The whole valley was worth saving, so that's what we decided to try."

The Lindbergh family backed up those words. In the early 1980s, they placed a conservation easement on three miles of Blackfoot River frontage—763 acres inside the Lindburgh Cattle Company, which is now part of the Paws Up Resort. Today, resort buildings crowd the edge of protection, but no farther. Land and Susie then donated an easement on 240 acres beside the water—a place they call The Homestead. More easements followed.

In the first decade after the passage of Montana's Open-Space Land and Voluntary Conservation Easement Act, the Blackfoot River was an epicenter of private land conservation. Indeed, the desire of private landowners within the ever-busier river recreation corridor was the major impetus for the passage of a statute enabling the use of the conservation easement tool. But, as was the case with most landscapes in Montana, public lands often comingled with private ownership. Parcels of state land, for example, bordered private lands that were on track for conservation easement protection. In one such case, Bill Potter of the E Bar L Guest Ranch informed The Nature Conservancy that the ranch wasn't willing to donate an easement unless the adjacent state land could be protected from subdivision and development. His position was understandable because the state had previously engaged in cabin site leases on government land right across the river from the E Bar L Guest Ranch.

The Nature Conservancy then pursued a strategy to get Fish, Wildlife and Parks to purchase a conservation easement from the State Lands Department, using dollars from Montana's Coal Severance Tax Trust Fund. The money was needed to compensate the State Lands Department, whose management mission was to generate income for Montana's State School Trust. This clever solution to protect public land within the river corridor was a bit of magic. It also created an "aha moment" in which The Nature Conservancy reasoned it would be a great idea if FWP could develop an in-house capacity to use conservation easements as an effective tool for wildlife habitat protection. Up to this point, the agency had used only deeded land purchases as a means of protection; it had accepted the gift of a small conservation easement on Sourdough Creek south of Bozeman, but it had no formal program for pursuing conservation easements as part of its habitat protection tool kit. FWP contracted TNC to help establish such a focus. Today, purchased conservation easements are the preferred alternative for the agency's habitat protection and public access objectives.

Conserving ground had begun, but recreational use by floaters and fishers was rising. Landowners faced increasing problems with trespassing, trash, and fires because the public believed that river access was a God-given right. It wasn't. Besides, the trout fishery remained damaged by overfishing, mining pollution, and clear-cut mud. In response, the Big Blackfoot Chapter of Trout Unlimited (TU) was formed to bring the fishery back. Roy O'Connor was a board director of the Montana Land Reliance back then. He remembers, "The riparian habitats weren't in great shape and bull trout were hard to find." J. Munroe McNulty donated $4,000 to TU to answer one question: "Why are there so few fish in the river?" He noticed that "there are few stoneflies left and caddis populations are sparse." Dennis Workman of FWP knew he was right. He said, "I caught and released a single cutthroat trout in the head of the Cottonwood Creek pool. I forded below the pool and fished it, using both wet and dry flies. I rested it and tried again with different flies. Not one damn fish."

As a result, "catch and release" and size regulations were established for the river. By 1991, a state fisheries biologist named Don Peters noted, "We're starting to see some nice improvements in adults and good records of larger fish." The number of 12-inch trout tripled to 150 per mile and tributaries began to heal. Spawners returned to Nevada and Blanchard Creeks. Elk Creek and others were restored.

Public use would be handled at state-owned fishing access sites and designated spots on private land—all part of Jerry Stokes's plan. Recreation Management Agreements were signed by public agencies and willing private landowners. Fishing and floating access points on ranches would stay open as long as visitors acted responsibly. The first agreements were for one year, with signs erected explaining the rules. "Walk-In-Only" hunting regulations were enforced by the FWP. It was an innovative gamble, and no one knew whether it would work. There was still so much land exposed to subdivision, and so many complex management issues left to solve. Land Lindbergh saw the need for strategic thinking: "Before, there was no forum to handle the impacts on the watershed. With the influx of new ideas and people, coupled with different agendas of all the agencies, it was time to get in front of potential issues and deal with them."

Opportunism was no longer enough.

The Blackfoot Challenge

In 1993, a group called the Blackfoot Challenge was formed. Rancher Jim Stone said, "We got tired of complaining about what we couldn't do, so we decided to start talking about what we *could* do." The group's strategic mission was "to coordinate efforts to conserve and enhance natural resources and the rural way of life in the Blackfoot watershed for present and future generations." All issues were on the table—development, timber harvest, restoration, weeds, grazing, water rights, fisheries, fire, soil, hunting, wildlife, and recreation. All stakeholders were invited to participate—ranchers, agencies, conservation NGOs, outfitters, businesspeople—everyone. Resource mapping was done, and a practical notion of sustainability drove things forward.

Conserving the Blackfoot came from connections, not conflicts. The tenets of the Blackfoot Challenge reveal its values:

- Be inclusive; invite everyone to the table.
- Identify community leaders who are respected and respectful.
- Practice the 80/20 rule. When you think of barbed wire, it's the pointy part that comes to mind, when in reality the majority of barbed wire is smooth. We believe it's the same when talking about values. If we focus on what we have in common, the values we share, we can get stuff done without getting hung up on the barbs.
- Be open, transparent, and honest. This builds trust and credibility.
- Do not pick sides or take positions on issues. This puts trust and credibility at risk.
- Practice "proper pacing." Stay in communication with your partners and don't make decisions without them.
- Facilitate a respectful conversation.
- Make decisions by consensus.
- Be willing to take it slow. Finding success through collaboration takes time. It will be worth it in the end.
- Get stuff done, share your successes, and celebrate!

The Blackfoot Challenge shined light on the future of the landscape by mirroring the "7 Rs" of Indian people—Respect, Relevancy, Reciprocity, Responsibility, Rights, Reconciliation, and Relationships. The Challenge became a clearinghouse for information and a unifying source of inspiration. Its primary job was and is "keeping the patient alive"—keeping development off key lands. Tax incentives for donated easements expanded and creative sources of money were found to buy conservation easements. Conservation easements in the Blackfoot now protect more than 200,000 acres.

Upstream from Lindbergh family easements at Greenough, nearly all land surrounding Ovando is conserved. That landscape of glacial hummocks, pothole lakes, and prairie will remain intact. Curving swaths of protection extend up Nevada Creek, safeguarding good ranch country. A remote stretch up Douglas Creek will produce grass, cattle, wildlife, and solitude forever. Up near Lincoln, land is protected in the Keep Cool and Poorman Creek drainages. Landscape ecology

North Fork, gateway to the Bob Marshall Wilderness

advises that connectivity is our best bet for maintaining natural resilience in the face of change. The Blackfoot watershed is a case study of how to get that done.

The Montana Land Reliance (MLR) website contains an interactive map of the state that allows you to click on an individual project to see who holds the easement and when it was completed. Former board codirector Roy O'Connor lives in the Blackfoot and helped create many of those easements, including one on his own property. MLR is a major force here and along other rivers like the Madison, Yellowstone, Smith, and Stillwater.

Writer and filmmaker Annick Smith donated a conservation easement to MLR on her 163-acre place near Potomac. Annick raised her family in an old mud-chinked cabin. In her book *Homestead*, she writes, "Our place sits on a meadow a mile and a half above the Blackfoot, but the river's presence ripples in my imagination as if it murmured outside my door. The Big Blackfoot has become my metaphor for change and connection." Her easement is fully tangible. As writer Bill Kittredge reminds us, "In a story, nothing is real until it is acted upon."

Montana Fish, Wildlife and Parks acquired conservation easements on lands it did not seek to purchase. Huge amounts of money were saved by using these less-than-fee-simple acquisitions. FWP easements now cover vast areas, including projects on state lands adjacent to the river and a 13,000-acre property on Chamberlain Creek near Sperry Grade.

The Nature Conservancy has a major presence protecting land along the Blackfoot and Clearwater Rivers. TNC keeps landowners involved, building confidence and lining up deals. In 1998, Bill and Betty Potter donated a 4,000-acre easement to TNC that allowed timber management by the Lubrecht Experimental Forest. This happened more than 20 years after the first easements in the valley. Bill was finally ready. He said, "You've got to leave behind something besides a mess in this world." His daughter Juanita Vero now runs the E Bar L as a guest ranch. In 2015, TNC purchased 117,000 acres of former Plum Creek timberland around Placid Lake, and the Gold Creek drainage next to the Rattlesnake Wilderness. Swaths of purple expanded across the map.

The Five Valleys Land Trust in Missoula holds easements from Potomac to Lincoln—a true expression of its geographic name. Much more on this important group is upcoming in the "Missoula Region" chapter.

The US Fish and Wildlife Service played a transformational role in the Blackfoot. The agency purchased conservation easements on thousands of acres in the watershed, a major focus in the state. Gary Neudecker of USFWS secured funding from the North American Wetlands Conservation Act, federal migratory bird programs, and the Land and Water Conservation Fund. From modest beginnings in the Blackfoot, land conservation was "scaling up"—going large and making history.

Mary Stranahan helped make that happen. She's a doctor who practiced on the Flathead Reservation and bought land in the Blackfoot near Marguerite Heller's spot, the retired schoolteacher. Mary is now the manager of Goodworks Ventures, LLC and part of the High Stakes Foundation, which grants money to nonprofit groups. Mary Stranahan is a lifelong giver, a healer, someone who serves. She placed a conservation easement on her land and then donated the property to the USFWS. The place now serves as the headquarters for agency operations in the watershed.

Jim Masar, a lawyer and county commissioner for Powell County, was a source of reliable information to landowners and helped reduce opposition to easements. It wasn't easy. Jim says, "In the early days it was tumultuous and ranchers wouldn't even say conservation easement or conservancy. But in time the opposition quieted."

So here we are—88 percent of the Blackfoot watershed is conserved in some way. But there's more to protecting the Blackfoot than easements and land purchases—management and restoration matter. The Blackfoot Challenge has a 29-member board including ranchers, outfitters, loggers, businesspeople, NGO staff, and agency folks. Seth Wilson is the executive director, overseeing a 10-person team. Seth has worked various Blackfoot Challenge jobs since 2001 and knows the ropes. The group's partners include 160 landowners, 30 businesses, 30 NGOs, and 20 public agencies. They view land protection as a noble first step in healing the Earth.

The Blackfoot Challenge helped create 27 fishing access sites along the river. These are a mix of state-owned and private sites where access agreements now run for 10-year terms. Signs declare this the Blackfoot River Recreation Corridor. The Blackfoot Challenge now manages the 41,000-acre Blackfoot Community Conservation Area, a mosaic of private and public land. It owns 5,600 acres in its heart. The group helps ranchers create water management plans. It also helps manage timber and use controlled burns to reduce fire risk and improve forest health. The US Forest Service thins small trees to lessen forest fires, and the Collaborative Forest Landscape Restoration Program extends that work across the Blackfoot, Swan, and Clearwater valleys. Weed pulls bring people together like Amish barn raisings. Norman Maclean would be impressed. In the late 1980s, he considered a last trip to the Blackfoot, writing, "It's a long way to come to view the clear cuts." But he added a prescient line: "Love has to bring us all back."

The Blackfoot *is* back, and so are trumpeter swans thanks to the Blackfoot Challenge and the USFWS. Meriwether Lewis saw trumpeters on the Clearwater River, but these birds were nearly extirpated in Montana when they were shot for their fashionable feathers. Today, populations exist in the Centennial Valley, Mission Valley, and for the past 20 years, in the Blackfoot. The bugle call of these elegant 25-pound birds is being heard again.

The Blackfoot Challenge also does less visible work. Wildlife carcasses are hauled off to reduce conflicts with grizzly bears and wolves. Stream flows are monitored and voluntary water conservation is achieved. Students are gathered to learn the real deal and pass it on. All this conservation began decades ago at the Sunset Hill School, a place for learning.

Jim Stone has been chair of the Blackfoot Challenge pretty much from the beginning. This rancher from Ovando turned a fine idea into a national model he calls "shared giving." He still runs the Rolling Stone Ranch with a wise hand. He even worked with the USFWS to restore a "fen wetland" on the property—a biodiverse peat bog, something scarce in Montana also seen in Pine Butte Swamp. Jim Stone is a champ, but as you've seen, the list is long.

It takes a valley to save a valley.

Connectivity

The Blackfoot is a Native place, traversed and revered for thousands of years. Without that stewardship, ecosystems would not have survived the mining and logging that followed. Much is made of Meriwether Lewis's journey through the watershed, but he was here for only three days. He explored an alien place, but at least he used the river's true name in his journal—Cokahlarishkit. Our tongues trip over the sound, but maybe the Salish and other Native people will teach us about that and so much more. Reconciliation occurs only between equals.

The Blackfoot Challenge website has a photograph of Hank Goetz and Land Lindbergh standing at the confluence of the Blackfoot and Clearwater Rivers. They wear blue jeans, caps, and Carhartt jackets. Hank and Land have been dear friends for 60 years through all that life throws at us. Dedication like that requires stamina, patience, and decency—traits perfectly suited to conservation. The photo glows with belonging.

Land describes it. "Hank and I talk about the physical calm that comes over us driving down Greenough Hill. We look down into the valley and up to those mountains. It's an amazing release. It's home. When you own a piece of ground here, it's like a marriage. You have a responsibility to tend it. There are so many ties and memories. It's the same for many of the easement givers in the valley. My kids feel it too." Hank chimes in: "You build a sense of belonging, a quest to belong to something far bigger than you are."

Or smaller. At the joining of those two rivers and two friends is a rare freshwater sponge named *Ephydatia cooperensis*. Cooperation. (Officially it's named for nearby Coopers Lake, but the word play feels truer.)

Jim Masar offers his take on why the Blackfoot was the scene of such success. "When money became available to buy conservation easements, that kind of turned the spigot on. Many ranchers became more amenable and so did the bankers they worked with. There's also a relatively small number of properties compared to some valleys. But it's more than that. These people go to church together, went to school together, are often related, and that's the chemistry of the whole thing."

Larch forest, Mission Mountains

The Blackfoot River flows through the core of Montana, connecting ecoregions and lives. It washes over ancient bedrock formed before there were fish. Trout are now diplomats of the river—a keystone species with a lot to say. So many heard that voice. Norman Maclean, his brother Paul, and son John fished together for the first time near the mouth of Belmont Creek. It's a place like many others on the river, but it's tough to drive by that reach without reflecting on the passage of time. Norman and Paul Maclean are gone now. So is Darrell Sall, the enlightened BLM director who helped create a 10-mile corridor upstream of Belmont. A memorial to Darrell is built just back from the river. Pull over next time and thank a friend you never met. Same for Ernie Corrick, an old-school timberman who saw a better future. Wet a line at Corrick's Riverbend Fishing Access Site and then rest in the willowy shade. Give thanks for the gifts of precious life, good people, and fine ground. Lend a hand, then celebrate.

Most of the Blackfoot is safeguarded, but for much of the twentieth century, it was abused and beaten. In time, we felt Nature's pulse and got to work. Former MLR codirector Roy O'Connor says of this rebirth, "It's pretty amazing how well she does given a chance to recover." Maybe honoring this special place means so much because it reminds us *we* can heal. Be restored and return the favor. And to do that we need to reach out to each other with compassion and respect—we need to connect.

Beyond all the techniques and capital required, that is why so much of this singular watershed was conserved.

FURTHER READING

Blackfoot Challenge. www.blackfootchallenge.org

"The Blackfoot River: A Landowner's Approach to Management." 1983. Hank Goetz. *Western Wildlands* 19, no. 1: 20–23.

Homestead. 1995. Annick Smith. Milkweed Editions.

Home Waters: A Chronicle of Family and a River. 2021. John N. Maclean. Mariner Books.

Montana Ghost Dance: Essays on Land and Life. 1998. John B. Wright. University of Texas Press (portions of this chapter are drawn from chap. 6, "The Real River That Runs through It").

Montana Land Reliance. www.mtlandreliance.com

A River Runs through It and Other Stories. 1976. Norman Maclean. University of Chicago Press.

CHAPTER 11

The Greater Yellowstone

Grizzly 399 is a celebrity around Yellowstone National Park. In May 2023, she emerged from hibernation with a playful cub at her side. A normal sight. However, biologists were stunned because Grizzly 399 was 27 years old, the longest-lived bear ever recorded around the park. Frank van Manen says she produced at least 18 cubs in her lifetime, an extremely high level of fecundity for a species known for slow reproduction rates. Frank runs the Yellowstone Interagency Grizzly Bear Study Team and recalls that 50 years ago there were only 136 griz within the entire Greater Yellowstone region covering 23 million acres. Today, there are about 1,000, with 200 spending most of their time in the park.

Grizzly 399 helps explain that recovery. She is far ranging, moving from the Teton Range north into Yellowstone and beyond. She tolerates human presence but never developed a taste for human food—picnics or people. That's true of most griz, but Grizzly 399 is a very public demonstration of that adaptive trait. Bear attacks make the news; peaceful encounters don't. In *Eight Bears*, Gloria Dickie explains our understandable fear: "The bear (griz) is more than 800 pounds of muscle and fat, ending in sharp canines and 4-inch claws. They are extremely defensive, ready to neutralize a perceived threat at a moment's notice. Such attributes dictate that the grizzly is the ultimate test of human acceptance."

That remains undecided, so bears rarely die of natural causes. Some are euthanized after killing cattle, others are shot in self-defense, and many others die of suspected human actions. Our fear of being eaten runs deeper than the facts. Since 1872, when the park was created, nine people have been killed by griz out of over 130 million visitors. Bear expert van Manen says, "Visitors to the heart of grizzly bear country in Yellowstone have about the same likelihood of being killed by a falling tree as being killed by a grizzly." However, outside the park, attacks sometimes happen as bears spread out from Yellowstone into a growing region. "How many bears is enough?" remains a valid question. Hannibal Anderson ranches in the Tom Miner Basin of the Paradise Valley. He says it this way: "I don't see the world as a place where humans just get to trump everything else. I consider it a fundamental responsibility of being human to serve the ecological integrity of wherever we live." In *Preserving Yellowstone's Natural Conditions*, James Pritchard substantiates the notion that park management has regional implications.

Grizzly 399 is a world-famous ambassador, seen by millions in photographs and videos. Viewers are amazed by the way she moves peacefully past cars and onlookers, carefully tending her cubs, sometimes three at a time. In 2020, there were four. "She's an icon, no doubt about that," said Dr. Chris Servheen. He ran the USFWS Grizzly Bear Recovery Program for 35 years and was a major force in getting 399 and her cousins protected by the Endangered Species Act. That status has mostly held through the years despite fear-based rhetoric and tough state politics. Importantly, healthy habitat exists for griz and hundreds of other species because of work done by the Montana Land Reliance, USFWS, National Park Service, The Nature Conservancy, Greater Yellowstone Coalition, Gallatin Valley Land Trust, and others. This vast landscape has been called an ecosystem based on the needs of far-ranging species like *Ursus arctos horribilis*—the grizzly bear. That understanding formed in the late 1950s when two young biologists studied bears inside the park.

Frank and John Craighead were twins who shared a passion for learning from wildlife. Many of us grew up watching their National Geographic Society documentaries. The Craigheads were experienced field hands, but grizzlies presented unique problems. At the time, existing knowledge came mostly from observations during daylight

hours and little was known about their overall movements or denning locations. The Craigheads turned to Philco Corporation, an electronics company, to brainstorm a way to track these nomadic bears. A technological solution was devised: after an animal was tranquilized, a radio collar was placed around its neck and colored tags attached to its ears. Identifying and tracking bears became much easier with this early form of GPS. Reams of fresh data opened the eyes of the Craighead brothers and the National Park Service.

It turned out that grizzlies, especially males, had ranges of up to 600 square miles. Telemetry showed that griz wandered wherever they liked, paying no attention to jurisdictional boundaries. They could no longer be thought of as "Yellowstone bears," but as part of something the Craigheads called the "Greater Yellowstone Ecosystem." This concept shifted our consciousness away from narrow land management to bioregional awareness. In the 1980s, Yellowstone superintendent John Townsley and well-regarded environmentalist Rick Reese discussed the need to address issues that threatened the park's integrity from outside its borders. At the time, government agencies and environmental organizations were "siloed," with little coordination of their activities. Rick Reese held numerous gatherings to see what could be done. In 1983, the Greater Yellowstone Coalition was formed with Reese as its first president. In 1989, he published a groundbreaking book called *Greater Yellowstone: The National Park and Adjacent Wildlands*. Today, this perspective is widely embraced by scores of agencies and NGOs.

However, the Greater Yellowstone concept predates all this by millennia. Shane Doyle says that 26 Indian tribes have formally recognized their connections to this ecosystem. Indigenous presence goes back at least 12,000 years. Brandon Sazue (Crow Creek Sioux) said, "In our genesis, it was the great grizzly that taught the people the ability to perform healing and curing practices, so the grizzly is perceived as the first 'medicine person.'" Many tribes moved up to what is now Yellowstone Park during warm months to hunt, fish, collect plants, and pursue spiritual practices. As winter approached, tribes moved down in elevation, then across the region to core homelands within the Greater Yellowstone. This mirrored the movements of many bison, elk, deer, and other species that returned to low-elevation winter ranges. The Greater Yellowstone is an ancient cultural and ecological reality.

OPPOSITE: Bison ford, Gardner River

The Greater Yellowstone Area

Such a vast expanse. Yellowstone National Park, the world's first, forms the core, with public lands and ranch country surrounding it. About 64 percent of the Greater Yellowstone is public, 6 percent is tribal, and the remaining 30 percent is privately owned. There are seven national forests, three national wildlife refuges, three states, and 20 counties. In Montana these include Beaverhead, Carbon, Gallatin, Madison, Park, Stillwater, and Sweet Grass Counties. Most conservation groups work on public land issues such as wilderness, logging, wildlife refuges, mining impacts, and federal species protection. The Greater Yellowstone Coalition (GYC) is a prominent example that envisions "a Greater Yellowstone where wild nature flourishes; plant, animal, and human communities thrive in reciprocity." The coalition works to craft elk occupancy agreements with ranchers, create a Yellowstone cutthroat trout refuge, elevate tribal interests, stop mines on the park border, prevent wildlife conflicts, protect "iconic species," and defend wild rivers. GYC has a staff of 31. Mike Clark ran the organization for many years, leading it into prominence with a commitment to working across the political spectrum. He stressed private land conservation and led a successful effort to prevent a gold mine near Cooke City that endangered Yellowstone Park. GYC is science-based and collaborative, but it enters the political fray when it believes it's necessary. Its approach is substantially different from that of land trust groups, which focus mostly on securing conservation easements, trails, and parks.

Data show that two trends emerged on private land in the Greater Yellowstone between 2005 and 2018: the number of properties 640 acres or larger fell by 6 percent, and the average size of parcels increased. Fewer, larger holdings were juxtaposed

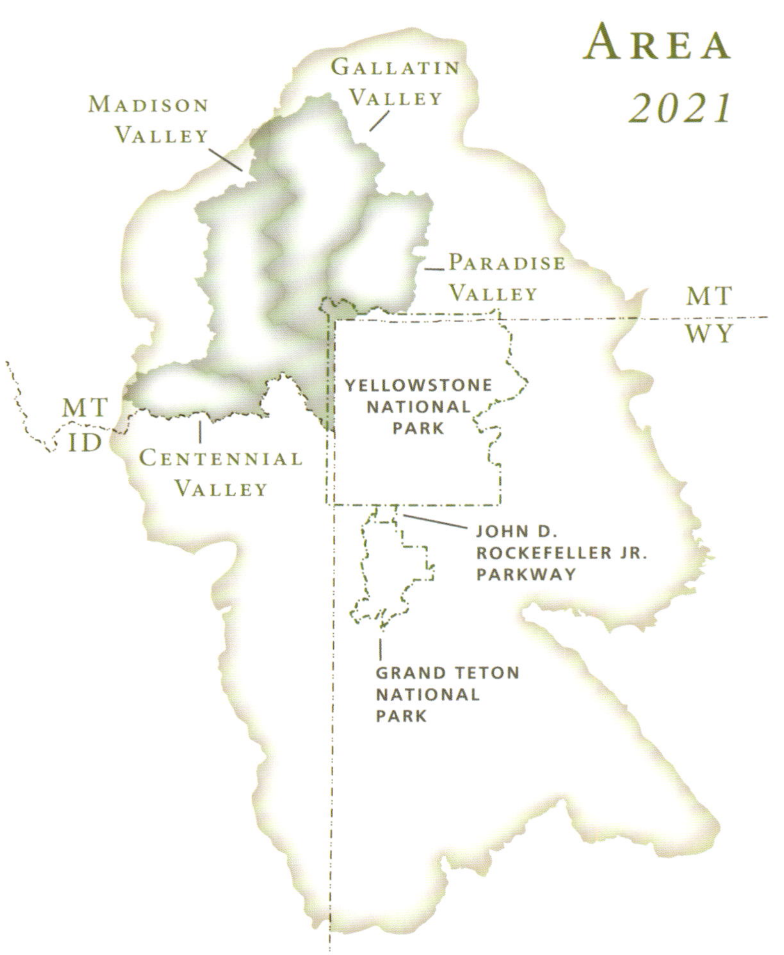

with an increase in small residential lots. Land was being both assembled into larger ranches and split into subdivisions. The first trend provides an opportunity for conserving, and the second shows why it is needed. *Saving the Big Sky* focuses mostly on private land conservation, and in this chapter we look at four scenic Montana valleys adjacent to the park—the Paradise, Madison, Gallatin, and Centennial. Northeast of the park, another 154,000 acres of MLR conservation easements are found in Carbon, Stillwater, and Sweet Grass Counties. But before we explore the Greater Yellowstone in detail, additional perspective is needed on why we emphasize private land here and throughout the book.

Paradise Valley, summer morning

Yellowstone River oxbow

The Role of Private Land in Conservation

by Todd Wilkinson, journalist

Yellowstone National Park serves as a high-profile, globally renowned beating heart of wildness, but it is the sum of everything around it. The 23-million-acre Greater Yellowstone extends from the Big Belt and Crazy Mountains in Montana hundreds of miles southward through Yellowstone and Grand Teton National Parks to the very tip of the Wind River Range in Wyoming. That southern extent is near the Red Desert, where another mammalian celebrity is being tracked.

Maybe you've heard of mule deer 255? She holds the record for the longest-known seasonal commute of her species in the American West. Radio-collared in 2016 by scientists affiliated with the Wyoming Migration Initiative, 255 was documented moving 242 miles one way between the Red Desert and the forested environs of Island Park, Idaho, not far from Montana's breathtaking Centennial and Madison Valleys. As part of her circumnavigation, she made it back to where she started. While the total distance she covered is itself remarkable—as are the similar epic treks she completed over consecutive years—it is truly heart palpitating to consider the perilous obstacles she faced. These included formidable topography, blizzards, and drought. She also faced a gauntlet of predators and human hunters, busy highways, natural gas development, and more. Despite it all, she survived.

The movement of ungulates via corridors is a conveyor belt of biomass—calories on the hoof—that fuels an unsurpassed terrestrial food web. The body of a mule deer or a one-ton bison is formed by grass, water, sunlight, and soil. When an animal dies in the wild it feeds a wide range of other creatures, and when they die, others eat them. All that decomposition adds nutrients back into the soil, spurring plant and tree growth that stores carbon dioxide.

Dr. Matthew Kauffman is a federal researcher with the US Geological Survey who oversees the migration initiative based at the University of Wyoming. He believes that 255's journey is yet another compelling example of what it means, in the modern world, to have a healthy, still-intact wildlife ecosystem. Further, he says that mule deer 255 likely could not have accomplished her feat if the ground she covered had been transformed into suburban and exurban sprawl. David Pak, a wildlife researcher, agrees. He predicts that if residential sprawl continues at the same pace as in the past four decades, mule deer numbers could be in trouble. Like us, wildlife use their own versions of superhighways and back roads to travel, but most of their natural road maps have been largely invisible or little understood.

Wild Migrations: Atlas of Wyoming Ungulates helps correct that. It is the first comprehensive analysis of long-distance migrations of elk, mule deer, moose, pronghorn, bighorn sheep, and mountain goats in the American West.

Based on this kind of research, Dr. William Rudd and others have demonstrated that parks and other protected areas are not large enough to maintain stable populations of species that need plenty of room to roam. When large, intact landscapes break apart, islands of habitat are left behind in their wake. And in these piecemeal pockets, species are at higher risk of disappearing.

I am a journalist who has written about conservation in the Greater Yellowstone and beyond for nearly 40 years. I've heard the following point repeated over the years with a deepening sense of urgency: fragmented thinking and fragmented approaches to land protection result in fragmented landscapes. It takes patience for people to complete projects that coalesce into landscape-level conservation.

As capacious as the Greater Yellowstone's tapestry of national parks, national forests, and other public lands is, keeping private lands protected—not converted to subdivisions—is pivotal to saving the integrity of the whole. The ecological and economic future of the region depends on it. The Wild Livelihoods Business Coalition is based in the Paradise Valley. Its website states, "Every business owner near Yellowstone knows that many of our customers are here to see the Park. Wildlife watching is the #1 activity those visitors come to experience." Right now, all those species are able to migrate or move, some adhering to ancient passageways over long distances. Mule deer, elk, and pronghorn have some of the lengthiest migrations. Grizzly bears, wolf packs, mountain lions, wolverines, and Canada lynx have home ranges covering hundreds of square miles.

Dr. Arthur Middleton of the University of California, Berkeley, compares elk migration to lungs. In April, a dozen different herds from three states converge on Yellowstone as the park "breathes in" elk following waves of greening grass. In autumn, the park "exhales" elk departing for lower-elevation winter ranges, often on private land. In the Madison Valley, as many as 10,000 elk arrive to spend the cold months. Growth pressure in this and other valleys is rising as Bozeman gains national attention. Beyond elk, about 80 percent of all mammals and birds found in the Greater Yellowstone depend on private land for their survival.

Working lands, be they ranches and farms, or tracts of undeveloped open space, are integral to overall ecosystem health. Conserved linkage zones between public lands and crucial winter ranges allow wildlife mothers to raise their young in peace. Keeping a ranch or farm intact greatly improves the odds of wildlife persisting. Agriculture can adapt to the needs of wildlife, but subdivisions cannot. Each development adds to a dangerous labyrinth of traffic, barking dogs, smells, fences, and conflicts in formerly intact valleys.

The late Brian Kahn was a successful conservationist who hosted a popular radio program on Montana Public Radio. He once told me, "Mom and Pop ranchers and farmers are the unsung heroes of our state. They've been the keepers of our rural heritage and the pastoral character of their lands. That is central to the image of Montana that is beloved around the world. Too often, we take their contributions to wildlife conservation for granted." In my book *Ripple Effects: How to Save Yellowstone and America's Most Iconic Wildlife Ecosystem*, I reported that more than a million ungulates in the region rely on private lands. The migrations within the Greater Yellowstone distribute a sense of place far and wide. They also rebind Indigenous people with lost or stored transcendent memory.

The threads holding this landscape together, however, are remarkably fragile. David Hallac, formerly chief scientist in Yellowstone, says that environmental threats to natural landscapes are "death by 1,000 cuts." In 2012, Dr. Charles Schwartz, then head of the renowned Yellowstone Interagency Grizzly Bear Study Team, was the lead author in an analysis of the impacts of rural development on wildlife. He concluded, "The negative environmental consequences of rural land development, including landscape fragmentation, have been widespread and extensive in the USA. Over 90% of land in the Lower 48 states has been logged, plowed, mined, paved or otherwise modified from pre-settlement conditions." Greater Yellowstone is so rare because its geographic remoteness spared it from succumbing to those effects. But in-migration of people fueled by the recent Covid pandemic and the rise of remote work has weakened Montana's sense of isolation. Beyond ecosystem integrity, tragically, we are losing the stands of rural communities and their collective knowledge of and care for the land. Back in the 1970s, a land trust was born to deal with that.

The Montana Land Reliance

Christine Torgrimson grew up watching the Bitterroot Valley be subdivided and developed. She feared that farmers and ranchers couldn't stay on the land unless something was done, not just in the Bitterroot but in places like the Greater Yellowstone. As a college student, Christine mapped and analyzed land divisions in the state, and then she joined the successful effort to pass Montana's conservation easement law. Soon, a young Californian named Barbara Rusmore arrived with experience in private land protection in Silicon Valley. The women joined forces and in 1978, the Montana Land Reliance was born as a statewide land trust.

Christine and Barbara assembled a team of dedicated conservationists including Bill Milton, Chase Hibbard, and Jon Roush—business

leaders, ranchers, and scholars. An economist named Bill Long gave MLR sound financial guidance on how to raise, spend, and leverage money. Allen Bjergo brought Agricultural Extension Service experience to the mix. Chase Hibbard was a prominent rancher who saw land conservation this way: "Don't legislate it, don't force it. Provide landowners the right tools, and if it's right for them, it will fall into place." Indeed, it did. As of this writing, the Montana Land Reliance holds conservation easements on over 1.3 million acres around the state, the most of any statewide or regional land trust in America. Large parts of that are within the Greater Yellowstone in Montana. Here's why.

In 1982, Bill Long and Bill Dunham of MLR fished the Madison River with a Wall Street executive named Herb Wellington. He owned the Longhorn Ranch beside the water and adored trout. Bill Long remembers, "MLR was not the only organization wanting to work with Herb. A national group ran hard after him, but for some reason he picked us. We were new and didn't have it figured out yet, but Herb believed in us." His easement was filed and his judgment turned out to be prescient. The Montana Land Reliance grew exponentially based on delivering a singular product—the conservation easement.

The film *A River Runs through It* made the value of riverside ranches skyrocket in Montana. People wanted to buy property, fish, and live a rural dream. Between 1978 and 1989, MLR completed 27 easements. After the movie came out in 1992, it finished 414, averaging 38 per year. Rock Ringling, Lois Delger-DeMars, Jan Konigsberg, Amy Eaton-Royer, Chris Montague, and other pros drove the work to even higher levels. In mid-1992, Ringling set MLR's sights on conserving one million acres. "We needed a goal. Nobody thought it was necessarily attainable at the time, but it gave us something to focus on. We figured that every acre under conservation easement in our lifetime was the measure of our job." "Cows Not Condos" was a slogan that galvanized their work across the state.

Paul "Rock" Ringling of MLR is always willing to talk about those achievements. He had a long career codirecting that land trust. Ringling remembers, "Back in the 1980s, a lot of smaller ranch properties became uneconomical. They had natural amenities, yet were about to be sold at a discount. This coincided with an explosion of wealth in the country and a whole different set of people started buying land in Montana; ones who didn't have to make a living from a herd of 150 cows. They bought for the fishing and open space, then put a conservation easement on the ranch." Ringling comes from an ag background in Montana, so he could easily talk to rural landowners about options.

When asked what else makes easements work in Montana, Rock Ringling smiles. "Trout and funding. Me and Bill Long would cold-call someone from a company or foundation and say we wanted to talk about fishing. That opened many doors, then we'd take those clients on float trips to fish rivers like the Big Horn or Madison. One prime ranch had 1,800 acres split up into twenties, so some of our clients came up with the money to strip off the subdivision and place an easement on the place. That led to another one nearby and it grew from there. When we started out, I tossed some Cheerios on a map as a way to get started. No big plan. In time you look up and you've got linkage and landscape-scale conservation. But you cannot 'target' a ranch or valley for easements. You should *never* do this, it would turn out horrible. You anger landowners that way and I wouldn't blame them. You have to show respect or you're out of business."

The Land Reliance, as it's often called, is a go-to land trust based on an extraordinary track record. Its book *A Million Acres* provides elegant writing and striking photographs about its success. MLR's website contains a deeper description of this history and an interactive map showing all easements in the Greater Yellowstone and across Montana. This NGO holds 298 easements comprising 324,240 acres of critical private land near the park. Beyond the four valleys we will soon discuss, this includes Carbon County (29 easements, 46,826 acres), Stillwater County (33 easements, 47,170 acres), and Sweet Grass County (50 easements, 59,547 acres). This happened because MLR and other groups realized that people were drawn by the allure of trout, and every property was a candidate for a conservation easement. They were right.

The Madison Valley

In 1970, this place was largely unprotected except for national forest lands, which were being intensively logged. From Hebgen Lake to Three Forks, the Madison was fine ranch country, but there was no assurance it would stay that way. The 2021 map shows the transformation—80 percent of the landscape is conserved in some way (with 20 percent of the watershed being private lands under easement). That's up from 52 percent in 1970. The major pattern is the point, but a few details suggest the scale of this achievement. Near Cliff Lake, the 8,433-acre Three Dollar Ranch is conserved in an extensively subdivided area along Highway 287. It lies north of an 11,163-acre Sun Ranch Partners easement with MLR, which is near a 6,876-acre easement donated by the same owner to The Nature Conservancy. Turner Enterprises donated a conservation easement to TNC on 62,569 acres, part of the Flying D Ranch total of 114,000 secured acres in two watersheds. As the Madison River flows north, easements become more common than unprotected land. Ennis is a trout town, and conserved ranches define the place. You often see people standing in the street trying out a new fly rod before floating a reach of the Madison. Their goal is rainbow, brown, cutthroat, bull, and golden trout. Fresh water from Yellowstone Park and tributaries flowing from the Lee Metcalf Wilderness provide a good outing. Conservation and fishing are natural partners.

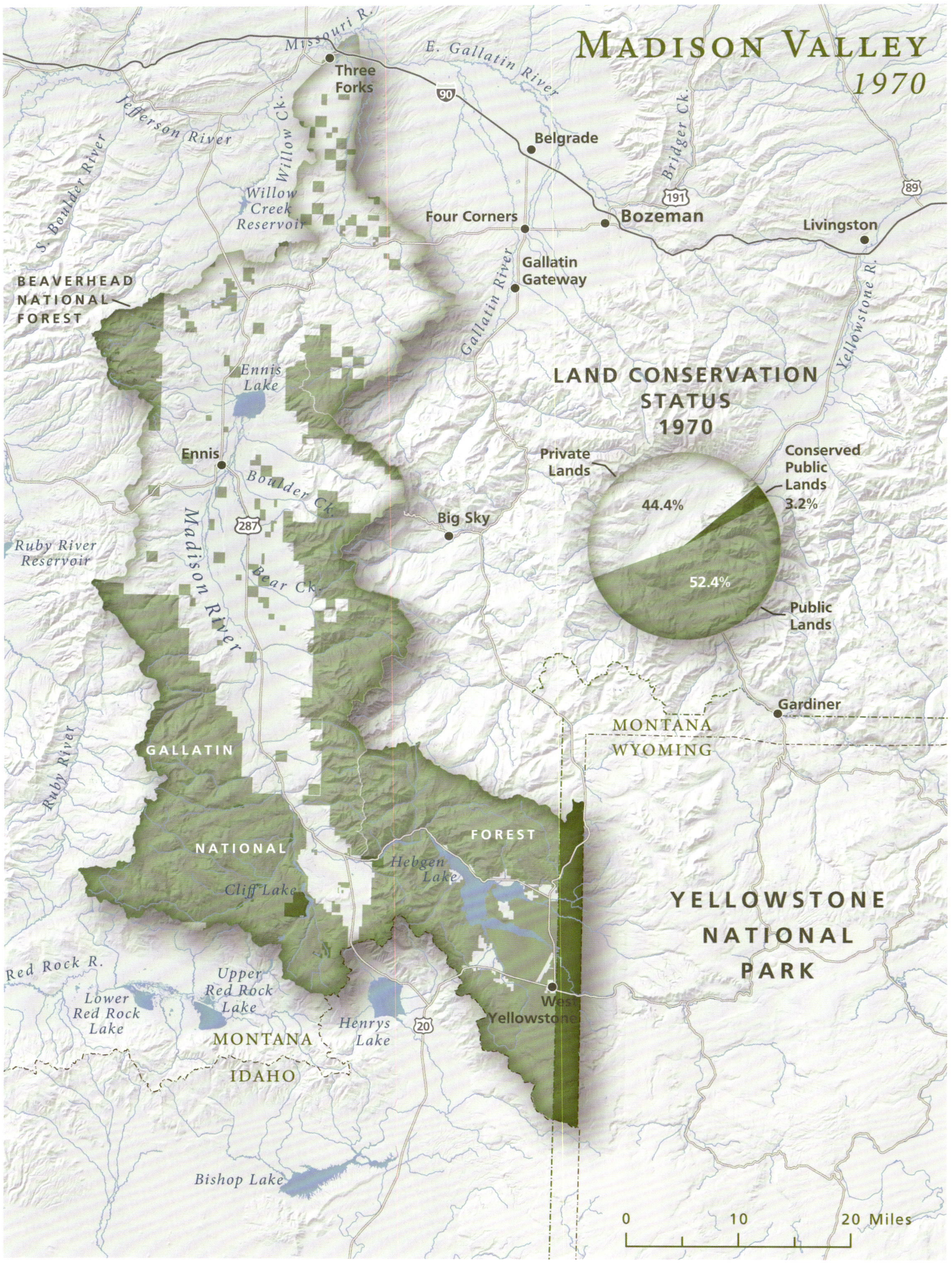

Madison Valley
1970
LAND CONSERVATION STATUS 1970
Private Lands
44.4%
Conserved Public Lands
3.2%
52.4%
Public Lands
Missouri R.
Three Forks
E. Gallatin River
90
Belgrade
Jefferson River
Willow Ck.
Bridger Ck.
S. Boulder River
Willow Creek Reservoir
191
89
Four Corners
Bozeman
Livingston
Gallatin Gateway
Gallatin River
Yellowstone R.
BEAVERHEAD NATIONAL FOREST
Ennis Lake
Ennis
Boulder Ck.
Big Sky
287
Ruby River Reservoir
Madison River
Bear Ck.
Gardiner
MONTANA
WYOMING
GALLATIN
Ruby River
NATIONAL
FOREST
Hebgen Lake
Cliff Lake
YELLOWSTONE NATIONAL PARK
Red Rock R.
Upper Red Rock Lake
Lower Red Rock Lake
Henrys Lake
20
West Yellowstone
MONTANA
IDAHO
Bishop Lake
0
10
20 Miles

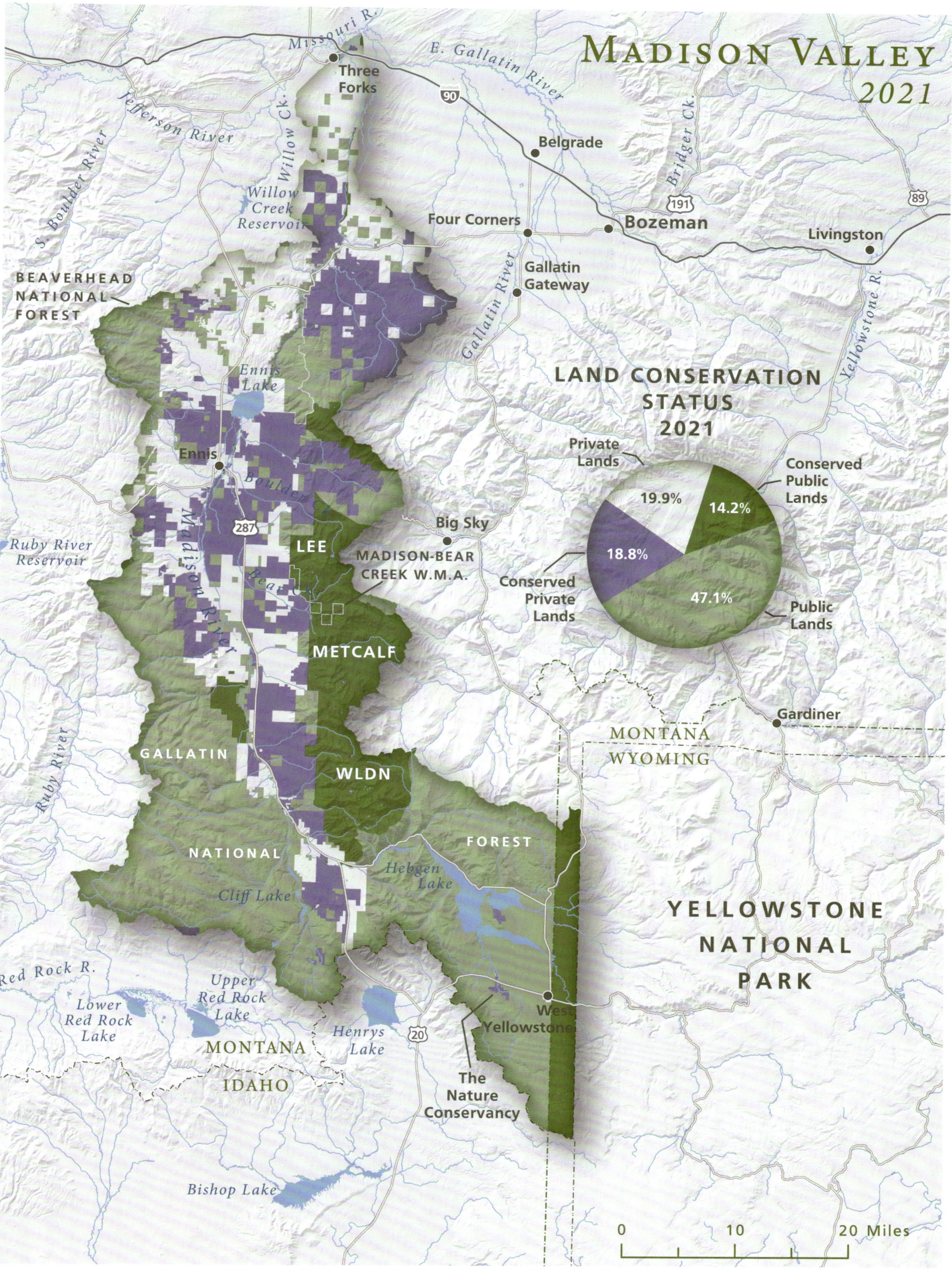

MADISON VALLEY
2021
LAND CONSERVATION STATUS 2021
Private Lands
19.9%
Conserved Public Lands
14.2%
Conserved Private Lands
18.8%
Public Lands
47.1%
Missouri R.
Three Forks
E. Gallatin River
90
Belgrade
Bridger Ck.
191
89
Bozeman
Livingston
Four Corners
Gallatin Gateway
Gallatin River
Yellowstone R.
Jefferson River
S. Boulder River
Willow Ck.
Willow Creek Reservoir
BEAVERHEAD NATIONAL FOREST
Ennis Lake
Ennis
Boulder
287
Madison River
Ruby River Reservoir
Big Sky
LEE
MADISON-BEAR CREEK W.M.A.
Bear
METCALF
GALLATIN
WLDN
NATIONAL
FOREST
Ruby River
Hebgen Lake
Cliff Lake
Gardiner
MONTANA
WYOMING
YELLOWSTONE NATIONAL PARK
Red Rock R.
Lower Red Rock Lake
Upper Red Rock Lake
MONTANA
IDAHO
Henrys Lake
20
West Yellowstone
The Nature Conservancy
Bishop Lake
0
10
20 Miles

The Paradise Valley

The Yellowstone River meanders through this stunning valley surrounded by high peaks in the Gallatin Range and the Absaroka-Beartooth. In 1970, no wilderness areas existed and ranches were vulnerable to subdivision. Many national forest lands were checkerboards with little management coherence. By 2021, conservation easements had secured large areas west of the river between Miner and Pine Creek. Forest Service holdings were consolidated and the Absaroka-Beartooth Wilderness was established in 1978. The Hyalite Porcupine Buffalo Horn Wilderness Study Area includes high country in the Gallatin Range. Over 15,000 acres of conservation easements exist near Corwin Springs, held by the US Forest Service. The Church Universal and Triumphant (CUT), a religious colony, owns much of the land encumbered by these easements. The Montana Land Reliance is very active in the valley north of here all the way to Miner and Chico Hot Springs. The Gallatin Valley Land Trust, a regional group, holds a large ranch easement near Pine Creek.

About 73 percent of the Paradise Valley is conserved, nearly doubling in 50 years.

This place is both true terrain and artistic heaven. Jim Harrison wrote *Legends of the Fall* here, then much more, punctuated by angling and storytelling at the Murray Hotel bar in Livingston. The pub also has a shrine to Václav Hazel, a Czech statesman, poet, playwright, and dissident. He would fit in. Tom McGuane fished, worked horses, and wrote endlessly in the valley, including screenplays for *Rancho Deluxe* and *The Missouri Breaks*, both authentically Montana. Richard Brautigan

Emigrant Peak

lived in Pine Creek and produced *Trout Fishing in America* (mostly not about fish), *The Hawkline Monster: A Gothic Western*, and *The Tokyo-Montana Express* (about salvaging a literary life when fame faded). Film director Sam Peckinpah lived out his last days at the Murray Hotel, pining for meaning until the end. Painter Russell Chatham was a friend to all of them, producing beautiful impressions of a valley they all loved. Paradise is a lot to live up to, but in art and land conservation, people have.

The Gallatin Valley

Bozeman is home to Montana State University and a world-class collection of conservationists. The Greater Yellowstone Coalition, Center for Large Landscape Conservation, American Rivers, Trust for Public Land, the Craighead Institute, and other groups are based here. The Big Sky Resort is an hour's drive up Gallatin Canyon, where skiing, hiking, and recreation abound. Yellowstone Park is only 80 miles away, making Bozeman an international destination. The city was named for the Bozeman Trail, a cattle route from Texas to Montana—*Lonesome Dove* country. But this community is now defined by scholarship, nature, tourism, high tech, agriculture, and trade. The downtown is full of great restaurants, bars, shops, and museums. No wonder growth is exploding here.

The Gallatin Valley Land Trust was created in 1990 with a mission to "connect people, communities, and open lands through conservation of working farms and ranches, healthy rivers, and wildlife habitats." Although based in Bozeman, GVLT works throughout the headwaters of the Missouri and other rivers. This includes the 4,390-acre Legacy Ranch and the Paradise Valley Ranch along the Yellowstone River, both outside the Gallatin Valley. Closer to home, easements hold agricultural landscapes together in the face of extraordinary development pressure. The Crawford property secures habitat for waterbirds, upland fowl, otter, and mink along the East Gallatin River. The Churchill area of the valley has the best agricultural soils in Montana. The Flikkema farm began as a homestead in the 1800s, so Maynard and Eileen Flikkema put their entire square mile under easement to secure a family legacy. The nearby Leep family farm will remain in grain production and perpetual open space. Sherwin Leep says, "This ground should never grow houses." Plainly said, and true for so many landowners who have worked with GVLT over the past three decades.

The group also partners with Gallatin County and the City of Bozeman. A county open space bond helps fund land purchases and easements. The Gallatin Valley Sensitive Lands Protection Plan forms partnerships between land-use planners, NGOs, and residents. GVLT also builds trails and creates publicly accessible open space. So far, 17 trails form a network across the area. The Main Street to the Mountains event celebrates biking and National Trails Day attracts community members to maintain beloved trails.

"Green infrastructure" is a foundation of Bozeman's beauty, prosperity, and health.

The Gallatin Valley Land Trust has worked with 116 families to conserve more than 50,000 acres. Its professional staff of 16 works across a vast region, and new easements are crafted month by month. Chet Work, the executive director, has 20 years of land trust experience, including leading the Teton Regional Land Trust. E. J. Forth, the associate director, says, "I am lucky beyond measure to live in such a special place. I take great comfort knowing that the work I do helps protect things I treasure most." The GVLT website provides hope based on accomplishments.

The Gallatin Valley is 66 percent conserved. The 1970 and 2021 maps illustrate a solid increase in land conservation over a half century. The Lee Metcalf Wilderness stands out on public land. It was named for a Montana senator from the Bitterroot Valley who was a staunch supporter of conservation in all its forms. A large purple area north of the wilderness is 42,000 acres of Turner Enterprises property secured by an easement held by The Nature Conservancy. Thousands of "eased" acres extend north of there to Manhattan and up the East Gallatin River. A 3,000-acre block stands out near Big Sky, land protected within the Yellowstone Club (a private resort).

The Museum of the Rockies is adjacent to the Montana State campus. Fossil dinosaur eggs and bones from the Rocky Mountain Front were

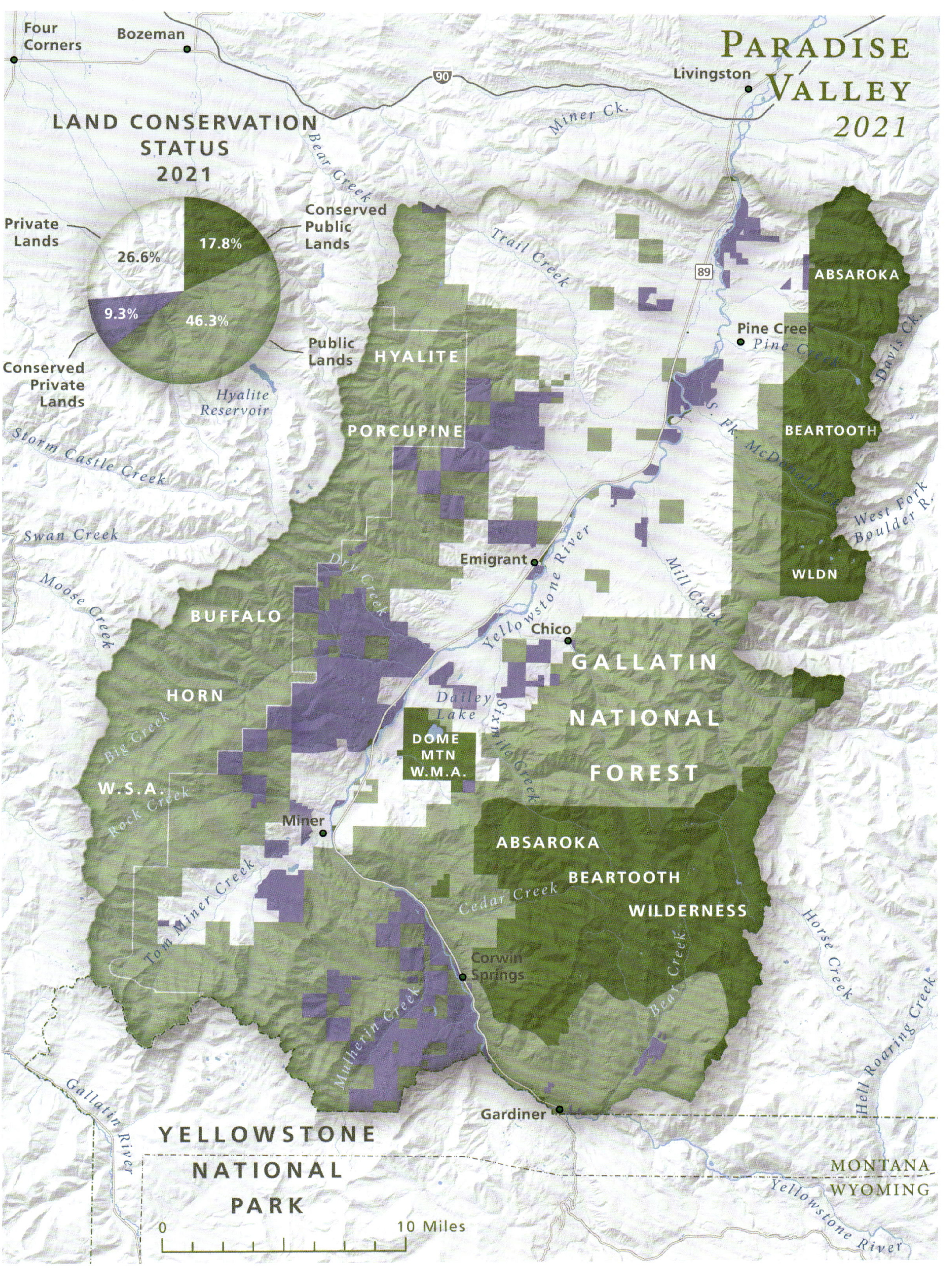

PARADISE VALLEY 2021
LAND CONSERVATION STATUS 2021
Private Lands
26.6%
17.8%
Conserved Public Lands
9.3%
46.3%
Public Lands
Conserved Private Lands
Four Corners
Bozeman
90
Livingston
Miner Ck.
Bear Creek
Trail Creek
89
Pine Creek
Pine Creek
Davis Ck.
ABSAROKA
BEARTOOTH
WLDN
West Fork Boulder R.
S. Fk. McDonald Ck.
HYALITE
PORCUPINE
Hyalite Reservoir
Storm Castle Creek
Swan Creek
Moose Creek
Dry Creek
Emigrant
Yellowstone River
Mill Creek
BUFFALO
HORN
W.S.A.
Big Creek
Rock Creek
Chico
GALLATIN
NATIONAL
FOREST
Dailey Lake
Sixmile Creek
DOME MTN W.M.A.
Miner
Tom Miner Creek
ABSAROKA
BEARTOOTH
WILDERNESS
Cedar Creek
Bear Creek
Horse Creek
Hell Roaring Creek
Corwin Springs
Mulherin Creek
Gardiner
Gallatin River
YELLOWSTONE
NATIONAL
PARK
0
10 Miles
MONTANA
WYOMING
Yellowstone River

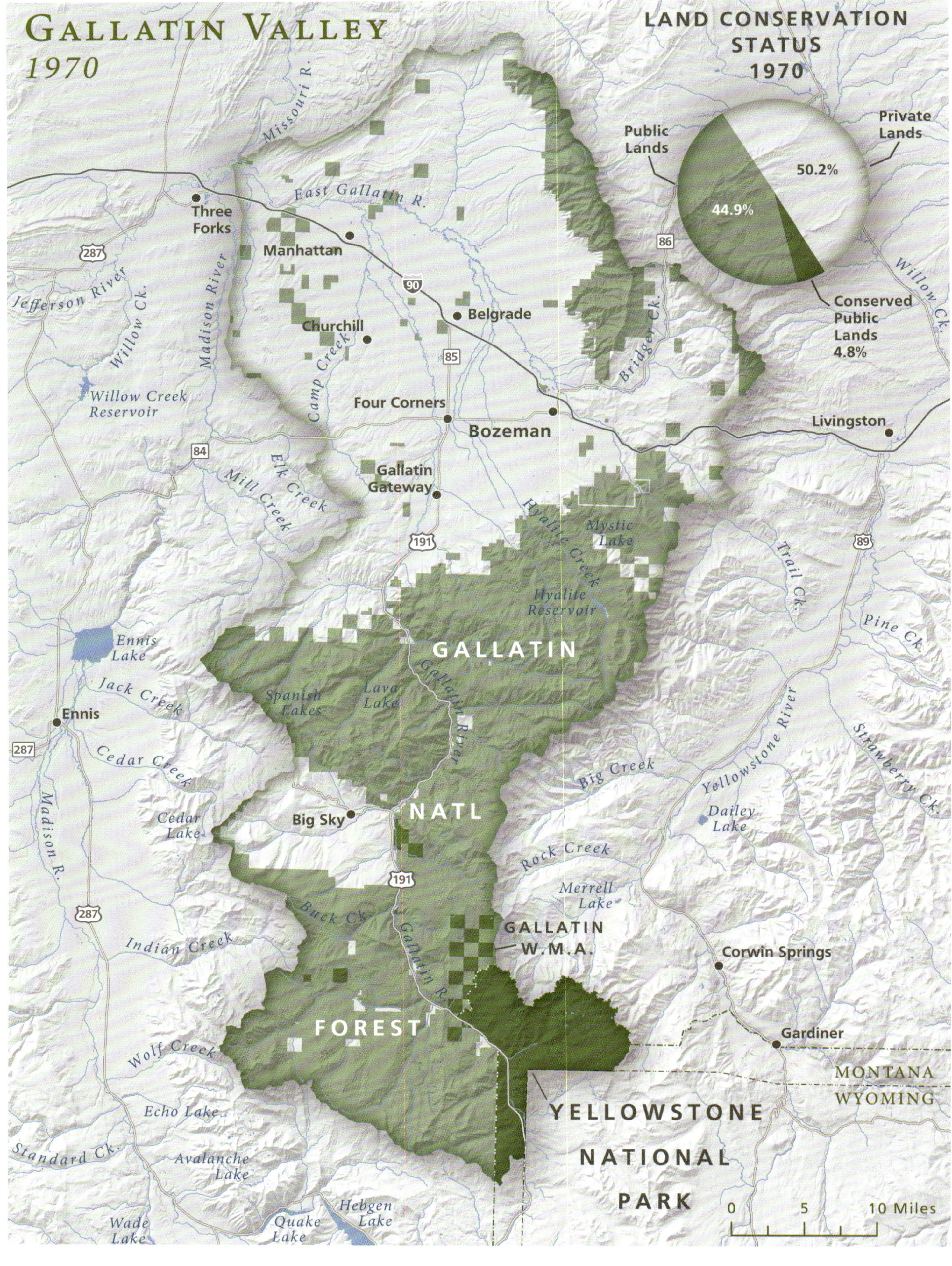
GALLATIN VALLEY
1970
LAND CONSERVATION STATUS 1970
Public Lands
Private Lands
50.2%
44.9%
Conserved Public Lands 4.8%
Missouri R.
East Gallatin R.
Three Forks
Manhattan
Belgrade
Churchill
Four Corners
Bozeman
Livingston
Gallatin Gateway
Jefferson River
Willow Ck.
Madison River
Camp Creek
Bridger Ck.
Willow Creek Reservoir
Elk Creek
Mill Creek
Hyalite Creek
Mystic Lake
Hyalite Reservoir
Trail Ck.
Pine Ck.
Ennis Lake
Ennis
Jack Creek
Cedar Creek
Cedar Lake
Madison R.
GALLATIN
NATL
FOREST
Spanish Lakes
Lava Lake
Gallatin River
Big Sky
Big Creek
Yellowstone River
Strawberry Ck.
Dailey Lake
Rock Creek
Merrell Lake
Buck Ck.
GALLATIN W.M.A.
Corwin Springs
Indian Creek
Gallatin R.
Gardiner
Wolf Creek
MONTANA
WYOMING
YELLOWSTONE NATIONAL PARK
Echo Lake
Standard Ck.
Avalanche Lake
Hebgen Lake
Quake Lake
Wade Lake
0 5 10 Miles
287
90
85
86
84
191
89

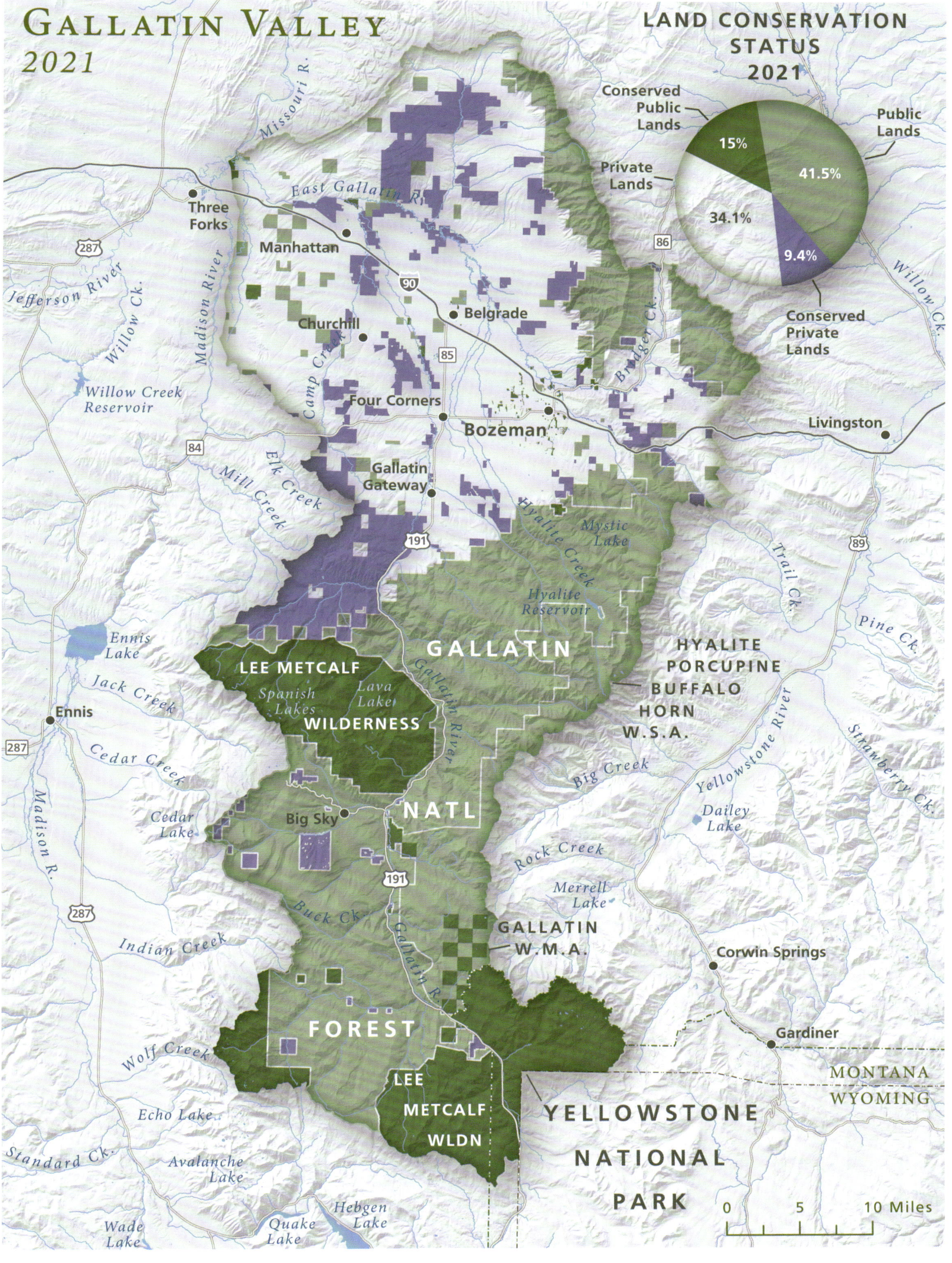
GALLATIN VALLEY
2021
LAND CONSERVATION STATUS 2021
Conserved Public Lands
15%
Public Lands
41.5%
Private Lands
34.1%
9.4%
Conserved Private Lands
Missouri R.
East Gallatin R.
Three Forks
Manhattan
Belgrade
Churchill
Four Corners
Bozeman
Livingston
Gallatin Gateway
Jefferson River
Willow Ck.
Madison River
Camp Creek
Bridger Ck.
Willow Creek Reservoir
Elk Creek
Mill Creek
Hyalite Creek
Mystic Lake
Hyalite Reservoir
Trail Ck.
Pine Ck.
Willow Ck.
Ennis Lake
Ennis
Jack Creek
Cedar Creek
Cedar Lake
Madison R.
GALLATIN
NATL
FOREST
LEE METCALF
WILDERNESS
Spanish Lakes
Lava Lake
Gallatin River
HYALITE PORCUPINE BUFFALO HORN W.S.A.
Big Creek
Yellowstone River
Strawberry Ck.
Dailey Lake
Big Sky
Rock Creek
Merrell Lake
Buck Ck.
Gallatin R.
GALLATIN W.M.A.
Corwin Springs
Gardiner
Indian Creek
Wolf Creek
LEE METCALF WLDN
MONTANA
WYOMING
YELLOWSTONE NATIONAL PARK
Echo Lake
Standard Ck.
Avalanche Lake
Wade Lake
Quake Lake
Hebgen Lake
0 5 10 Miles
90
85
86
84
191
89
287

brought here for study and safekeeping. Inside the museum is a T. rex skeleton found in 1988 near Fort Peck Reservoir in Eastern Montana. Dr. Jack Horner, a former curator of paleontology at the Museum of the Rockies, conducted the dig where fossilized blood was found. For good or ill, he was the inspiration for the *Jurassic Park* movies. With a kid's sense of humor, Jack put a replica of a T. rex in front of the building. He named it "Big Mike."

The Centennial Valley

"In a lovely valley surrounded by high mountains, are the Red Rock Lakes, which nature has designed especially for the swan," novelist E. B. White wrote a half century ago, and the place looks pretty much the same today. The 51,386-acre Red Rock Lakes National Wildlife Refuge is its ecological heart, and home to 300 trumpeter swans during the summer. In 1932 there were only 66, with the species making a last stand in these remote wetlands below the Gravelly Range and Centennial Mountains. This was sheep and cow country for families with the grit to endure this high, cold place. Wildlife were their closest neighbors: moose, mule deer, elk, pronghorn, grizzlies, lynx, and lots of frisky red foxes. All those species still flourish. This is true in part because many homesteaders gave up and sold their land, which became part of the wildlife refuge. Other ranch properties became conserved in recent decades. Biologist Chris Montgomery reports that "the Centennial has the densest population of peregrine falcons and ferruginous hawks in Montana as well as healthy populations of bald eagles and osprey. Westslope cutthroat trout thrive in Red Rock Creek."

And there are lots of trumpeter swans. This is the largest native waterfowl in North America, weighing more than 25 pounds, with a wingspan stretching six feet. They survived market hunting for feathers, habitat loss, and indifference until

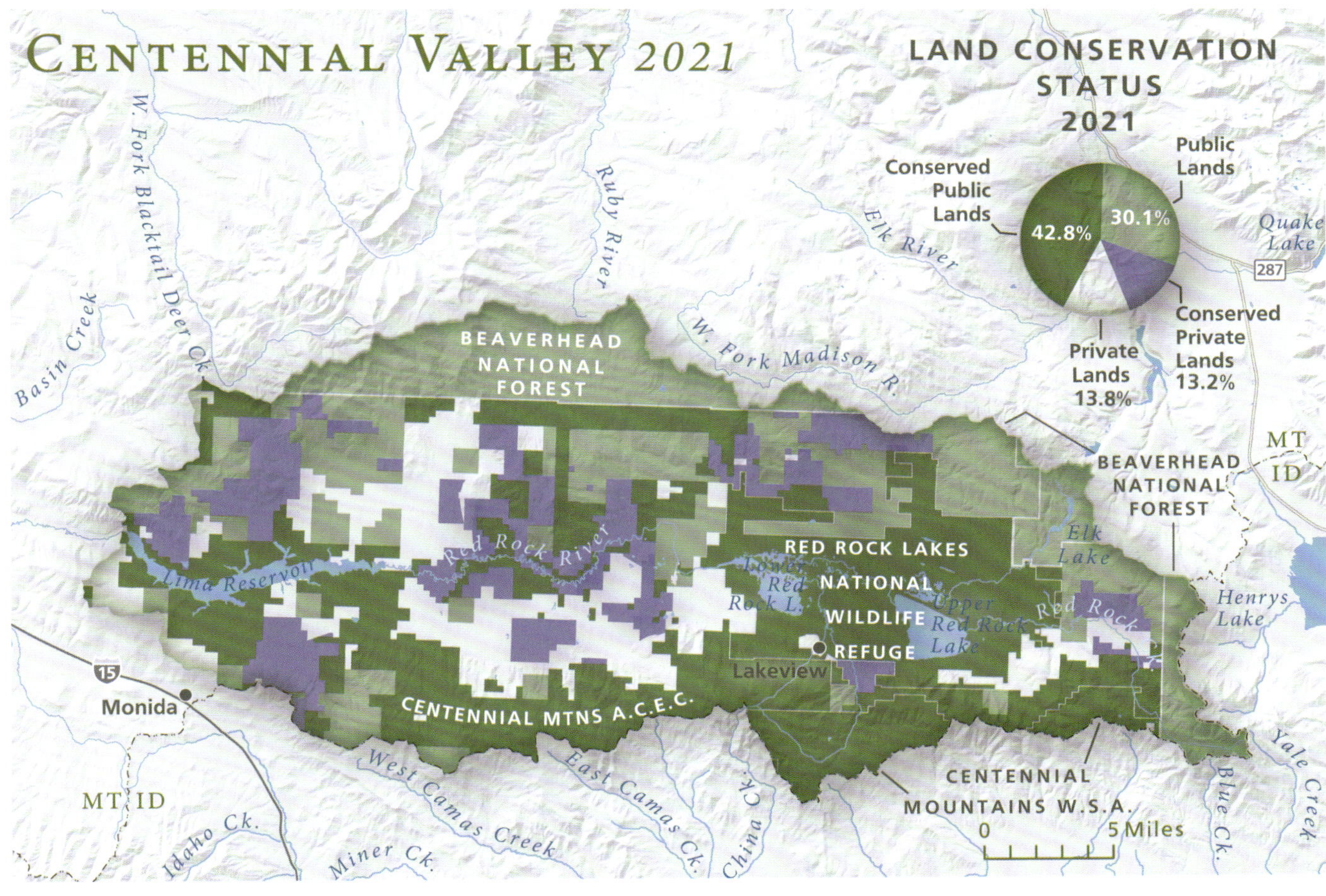

CENTENNIAL VALLEY 2021
LAND CONSERVATION STATUS 2021
Conserved Public Lands
42.8%
Public Lands
30.1%
Conserved Private Lands
13.2%
Private Lands
13.8%
W. Fork Blacktail Deer Ck.
Ruby River
Elk River
Quake Lake
287
Basin Creek
W. Fork Madison R.
BEAVERHEAD NATIONAL FOREST
BEAVERHEAD NATIONAL FOREST
MT
ID
Elk Lake
RED ROCK LAKES NATIONAL WILDLIFE REFUGE
Red Rock River
Lima Reservoir
Lower Red Rock L.
Upper Red Rock Lake
Red Rock
Henrys Lake
Lakeview
15
Monida
CENTENNIAL MTNS A.C.E.C.
CENTENNIAL MOUNTAINS W.S.A.
0
5 Miles
Yale Creek
Blue Ck.
MT ID
Idaho Ck.
West Camas Creek
Miner Ck.
East Camas Ck.
China Ck.

people cared. Swans are wary of people but fond of beavers, often nesting on top of their dams.

The Centennial is a fine sight in Montana, a spectacular landscape that is 86 percent conserved. The USFWS and The Nature Conservancy purchased land and secured easements across the valley. TNC manages the 1,400-acre Sandhills Preserve, where four rare plant species are found. On a macro scale, TNC says the Centennial "is a critical migration route of wildlife . . . a continuation of the High Divide Headwaters and the Big Hole. It maintains links between protected wildlands of Yellowstone, Central Idaho, the Crown of the Continent (Glacier), and Canada." Large mammals can adapt with enough good land to move into. The Centennial also has 241 species of birds in a place with a summer population of 100 humans. The avian family names exceed ours: greater sage-grouse, western grebe, cormorant, black-crowned night heron, white-faced ibis, sandhill crane, long-billed curlew, and waterbirds of all ancestries. In the winter, wildlife pretty much have the place to themselves.

But the Centennial used to be a major tourist route between Monida Pass and Yellowstone Park. The Utah and Northern Railway was Montana's first line, and between 1898 and 1913, tourists were dropped off at the pass and took a six-horse Concord stage through the Centennial. Horses were changed every 15 miles at a string of relay stations. Visitors had lunch at Lakeview, the halfway point, and then continued on to Yellowstone. Unlike in most of Montana, tourism did not lead to subdivision and housing development. The Centennial was too remote, buggy, and raw. Just to be sure, people stepped forward to protect the beauty and bounty forever.

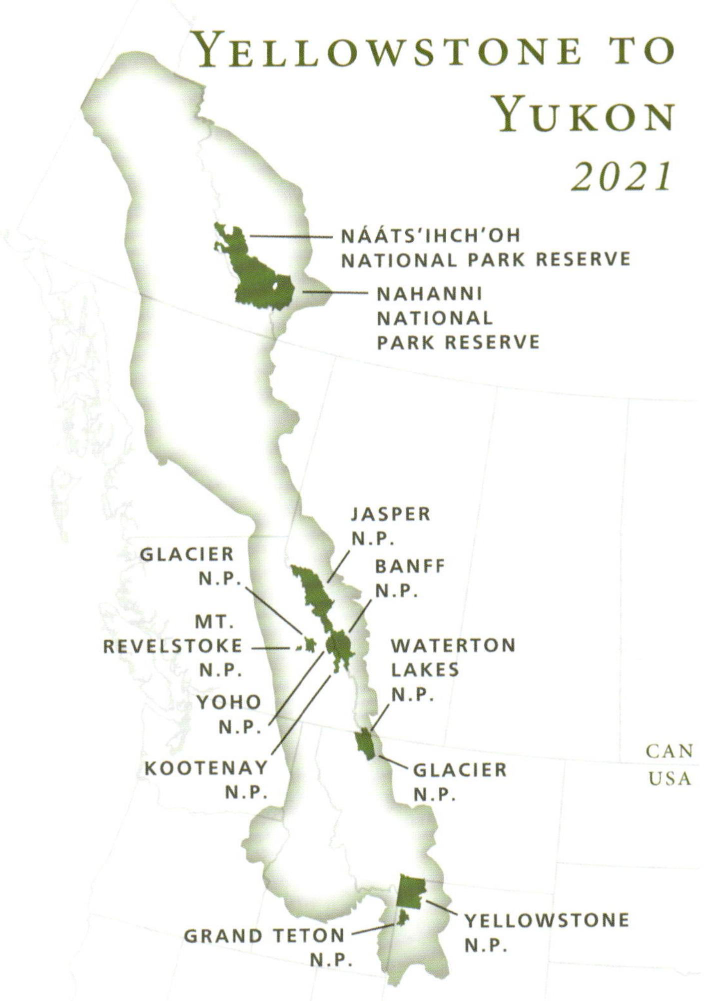

Yellowstone to Yukon Initiative (Y2Y)

In 1993, Harvey Locke had an epiphany. "A 14-day walking traverse of the Willmore Wilderness Park north of Jasper National Park, followed by a horse trip in the wild Northern Rockies of British Columbia made it clear to me that there is one gigantic linear ecosystem that extends from Yellowstone to the Yukon. By a campfire above Kelly Creek in the heart of what is now called the Muskwa-Kechika, I pulled out a pen and began writing on a topographic map. I wrote the words 'Yellowstone to Yukon' for the first time with the conviction that this was the right scale at which to think and act."

Harvey Locke went on to cofound the Y2Y Initiative as a joint Canadian-US nonprofit

organization. Its goal is to safeguard a 2,100-mile interconnected corridor extending north from Yellowstone National Park to the Yukon Territory of Canada. This "large landscape conservation" idea dovetails with the Spine of the Continent concept of biologist Michael Soulé. Dr. Jodi Hilty is the president and chief scientist at Y2Y, drawing on 25 years of managing landscape conservation projects. Her book *Corridor Ecology* is the gold standard on the topic. The Y2Y group focuses on "wildlands" but also stresses reciprocal relationships, "so people and nature can thrive." In the past 30 years there has been an 80 percent increase in conserved public land, mostly in Canada. Yet as we've seen, public land alone is insufficient to sustain landscape integrity. The four valleys we just explored clearly show why.

Conclusion

The television series *Yellowstone* is bunk. Ranchers in this region are not attacked by vicious developers, evil park rangers, and Indians acting like mob bosses. No motorcycle gangs invade private land. Enemies are not murdered and thrown into a canyon.

The real Yellowstone is a peaceful, ecologically vital region. Most land in the Montana portion of the Greater Yellowstone is conserved. As in so many places in Big Sky Country, the 30×30 idea has been exceeded voluntarily and with little anger. This suggests that conserving 30 percent of the world by 2030 is both possible and practical.

Journalist Tom Brokaw says this landscape "looks like Tibet," with wonders in all directions—expected things like great fishing, surprises like the sight of a wolf, joys like the sound of pelicans clacking their bills above you on a high peak. This magnificent cultural and ecological treasure is doing well thanks to the actions of NGOs, agencies, and private landowners. Ranching remains productive and real here, not the foolishness of Hollywood.

Except in one way. The Dutton family in *Yellowstone* resolved their disputes by placing a conservation easement on their ranch.

FURTHER READING

"The Centennial Valley." Chris Montgomery, Rick Graetz, and Susie Graetz. University of Montana. www.umt.edu (part of *This Is Montana*, an online series).

Corridor Ecology: Linking Landscapes for Biodiversity Conservation and Climate Adaptation. 2019. Jodi A. Hilty, Annika T. H. Keeley, William Z. Lidicker Jr., and Adina M. Merenlender. Island Press.

Eight Bears: Mythic Past and Imperiled Future. 2023. Gloria Dickie. W. W. Norton.

"Famous Jackson Hole Grizzly 399 Wows Again, but Now What?" May 18, 2023. Todd Wilkinson. *Mountain Journal*. www.mountainjournal.org.

"Fatal Grizzly Attack Renews Debate over How Many Bears Are Too Many." August 5, 2023. NPR *Weekend Edition*. www.npr.org.

Gallatin Valley Land Trust. www.gvlt.org.

Greater Yellowstone Climate Assessment. https://www.gyclimate.org.

Greater Yellowstone Coalition. www.greateryellowstone.org (the website illustrates the range and depth of the group's work).

Greater Yellowstone: The National Park and Adjacent Wildlands. 1991. Rick Reese. National Book Network.

Grizzly 399: The World's Most Famous Mother Bear. 2023. Tom Mangelsen and Todd Wilkinson. Rizzoli Press.

A Million Acres: Montana Writers Reflect on Land and Open Space. 2019. Keir Graff, ed. Alexis Bonogofsky, photographer. Riverbend. Produced by the Montana Land Reliance.

Montana Land Reliance. www.mtlandreliance.org (detailed information about their work).

An Origin Story: The Greater Yellowstone Ecosystem. September 23, 2020. Mike Koshmrl. *Jackson Hole News and Guide*. https://www.jhnewsandguide.com/special/conservation/an-origin-story-the-greater-yellowstone-ecosystem/article_825dae2a-odef-541a-8c43-c575cb4f8551.html.

Preserving Yellowstone's Natural Conditions: Science and Perception of Nature. 2022. 2nd ed. James A. Pritchard. University of Nebraska Press (the best analysis of park management issues and changing ideologies over time).

Radiotracking of Grizzly Bears. 1962 and 1963 Reports. Frank C. Craighead Jr. https://arc.lib.montana.edu/frank-craighead/objects/2650-010-010.pdf

Ripple Effects: How to Save Yellowstone and America's Most Iconic Wildlife Ecosystem. 2022. Todd Wilkinson. Mountain Journal.

"Rural Land Concentration and Protected Areas: Recent Trends from Montana and Greater Yellowstone." 2022. Julia H. Haggerty, Kathleen Epstein, Hannah Gosnell, Jackson Rose, and Michael Stone. *Society and Natural Resources* 35, no. 6: 692–700.

The Trumpet of the Swan. E. B. White. 1970. Harper & Row.

Wild Migrations: Atlas of Wyoming's Ungulates. 2018. Matthew J. Kauffman, James E. Meacham, Hall Sawyer, Alethea Y. Steingisser, William J. Rudd, and Emilene Ostlind. Oregon State University Press.

Yellowstone Interagency Grizzly Bear Committee. www.igbconline.org

Yellowstone to Yukon Conservation Initiative. www.y2y.net

Mount Jumbo in the middle ground, Rattlesnake National Recreation Area and Wilderness in the background

CHAPTER 12

The Missoula Region

The top of Mount Jumbo was once an island in Glacial Lake Missoula. Shorelines are etched into the prairie grasslands below, eroded by waves thousands of years ago. During the Ice Age, a massive glacier in Idaho advanced and blocked the Clark Fork River, which created a 3,000-square-mile lake. Over millennia, the ice dam was breached and re-formed dozens of times and the saddle of Mount Jumbo carried torrents of water as the lake drained. Boulders are strewn across the Rattlesnake Valley that don't match the geology below—erratics dropped by the deluge. This land is a witness to cycles of chaos and calm.

Standing on the summit of Mount Jumbo, you see Missoula spreading across the former lake bottom. In 1970, 30,000 people lived here—now 80,000 do. Word is out that Missoula is a great place to live, offering a university, a lively downtown, a big river, and endless open space to roam. As a result, development pressure has become extreme, part of something called the "Amenity Trap." The expansion of land conservation, cultural opportunities, and economic vitality attracts even more people, which generates greater challenges for affordable housing, transportation, and community services. Good places attract growth. Communities like Missoula, Bozeman, Helena, and Kalispell are shaped by the Amenity Trap, and each confronts the issue in different ways. Montana is not unique in the West. Boulder, Durango, Sante Fe, Eugene, and many other successful cities now must deal with the impacts of becoming magnets for growth.

From the top of Mount Jumbo, five major valleys are in view, each one a part of the story of Saving the Big Sky—Clark Fork, Rattlesnake, Blackfoot, Bitterroot, and Ninemile. Rock Creek is nearly visible to the east, a tributary thick with aquatic insects and famous trout. To hike from Jumbo to Glacier National Park, you cross only two paved roads. To trek from nearby Mount Dean Stone to Yellowstone National Park, you must "endure" three paved roads and an underpass of Interstate 15. Missoula is a hub of healthy rivers and immense wild terrain. This natural endowment also contains one of the largest open space systems in the country.

The Rattlesnake National Recreation Area and Wilderness rises north of Mount Jumbo, a kingdom of grizzlies, moose, and elk, Missoula's wild backyard. Rattlesnake Creek easements are bordered by North Hills easements and land acquisitions. Waterworks Hill harbors Missoula phlox, a whitish-blue endemic flower. The slopes of Mount Sentinel are protected by conservation easements and land purchases, including its famous trail to the *M* above the University of Montana campus. Forested Mount Dean Stone looms above it, newly acquired and safe. A linear park parallels the Clark Fork River, where you can bike to Milltown State Park, float from the Sha-Ron Fishing Access Site into town for a drink, watch surfers at Brennan's Wave, or stroll through a bountiful farmers' market full of locally produced food. Part of this trail system is named for Kim Williams, a renowned local NPR commentator and lovely soul. The Beartracks Bridge crosses the river where Chief Charlo (Claw of the Little Grizzly) and his band suffered a forced march from the Bitterroot Valley in 1891. Downriver is Council Grove State Park, where the Bitterroot and Clark Fork merge in a maze of cottonwood channels. In 1855, the Hellgate Treaty was signed here when tribal leaders felt compelled to cede most of Western Montana to the US government. It is a beautiful, heartbreaking place. Upriver is Milltown State Park (long a Salish campsite), where a dam was removed and copper smelter waste was sent back to Anaconda. The Clark Fork Coalition, based in Missoula, was the driving force to get that done. To the south is Travelers' Rest State Park, where Lewis and Clark camped on their way to the

Pacific and again on their return. Kiku Haines of The Conservation Fund secured the money to buy the site from ranchers. Farm and ranch easements extend west toward Frenchtown, protecting wheatfields and pastures. Farther on, easements in the Ninemile Valley keep agriculture intact and provide wolves a home. Way downriver on the Clark Fork, you can almost see mist rising from rapids in the forever-wild Alberton Gorge.

Imagine if these tens of thousands of acres had not been conserved. Fifty years ago, none of it was. Many residents and visitors enjoy this bountiful open space system with no idea of the work it took to create it. As you'll see, triumphs are never easy.

Five Valleys River Park Association

In 1972, 20 people gathered in Missoula to take on the impossible—establish a vast riverfront park system. People like Bob and Ellen Knight, a young lawyer and a biology grad student. University professors showed up like Arnie Bolle, Chris Field, Robert McKelvey, and George Babb, all committed to land protection. George Turman was there, a businessman who would later be the mayor of Missoula. Helen Bolle remembered their intention: "Our goal was to have river parks stretching from Milltown Dam to the confluence of the Clark Fork and Bitterroot Rivers." They knew it would take years to develop, so they got to work by engaging the community. It was a tough sell. At that time, the Clark Fork was severely polluted by mine wastes washing down from the copper smelter in Anaconda. Sometimes it ran red-orange through town. No one swam in the river and people were advised not to eat any fish they caught.

That same year, the Five Valleys River Park Association (FVRPA) got a meagre $100 seed grant from the America the Beautiful Challenge. There was no staff and no real money, so the group held community meetings, identified key lands, and advocated for their protection. The first completed project was Jacobs Island Park, next to the University of Montana. With Bob Knight's legal guidance, the Jacobs family bought this island in the Clark Fork and donated it to FVRPA. A Bureau of Outdoor Recreation grant spruced it up. Most of the island is now a dog park where poodles and mutts run free. Projects picked up from there.

Rancher D. J. Maclay donated a 10-acre riverside parcel to FVRPA. The group helped conserve 380 acres on Kelly Island, biodiverse riparian habitat where the Bitterroot and Clark Fork join. Tom Green Park was created on nine acres along Rattlesnake Creek. Henry Bugbee, a philosophy professor at the University of Montana, followed his ethics and completed a "bargain sale" of eight acres of bottomland forest nearby. FVRPA worked with The Nature Conservancy, Missoula County, and Montana Fish, Wildlife and Parks in using federal Land and Water Conservation Fund money to make this purchase. A boulder on the site has Professor Bugbee's words carved in deep: "In attending to this wilderness I knew myself to have been instructed for life." The deals grew more complex and more money was needed.

In 1980, Missoula voters passed a $500,000 open space bond. People agreed to pay slightly higher property taxes to fund land protection. An inventory of conservation resources was assembled with maps showing "keystone" properties such as river corridors, Mount Jumbo, Mount Sentinel, Blue Mountain, and the Rattlesnake Creek watershed. An Open Space Advisory Committee green-lit three projects. First, a stretch of former Milwaukee Railroad right-of-way was purchased in downtown Missoula on the south side of the Clark Fork. This is now a vital section of the riverfront trail system that includes John Toole Park, named for a prominent businessman and conservationist. Second, a 300-acre conservation easement was purchased on the face of Mount Sentinel, property owned by the Cox family. This was one of the first easements in the Missoula area. Finally, the city bought 125 acres north of the *L* on Mount Jumbo, but most of the regional terrain was still unprotected. The bond was spent and new directions were needed.

What follows is the interwoven chronicle of major land conservation achievements in Missoula and the Five Valleys region. Key facts and quotes come from Greg Tollefson's excellent write-up, "A Gracious Civic Enterprise: Five Valleys Land Trust History, 1972–2022." Timelines and places crisscross, since that's the intricate nature of this work.

The Rattlesnake

A deep pool glimmers in the sun. Argillite gives the illusion the water is green, but as the creek rushes along it reflects every ambient color—red osier dogwood, sky blue, yellow cinquefoil. The watershed was never extensively logged, so Rattlesnake Creek flows transparently clear. Houses were never built in the upper watershed, other than a few failed homesteads. This magic pool is only a 20-minute walk into the Rattlesnake National Recreation Area where people and dogs wade, but no one fishes. Visitor use is too high to allow even catch-and-release trout angling except above Beeskove Creek, six miles north of the entrance. The main trail leads upstream 15 miles until you enter the Rattlesnake Wilderness, and beyond that, cirques hold pure glacial lakes. Grizzlies live up there and den down here along Spring Creek. People hike, bike, or ride horses, exchanging greetings as they pass. They depend on this 61,000-acre reserve for exercise, solace, beauty, and the ability to live a natural life. It is a cornerstone of the Missoula community, but there was a real chance it could have been lost.

In the late 1970s, the Montana Power Company (MPC) owned 21,000 acres in a checkerboard with Lolo National Forest lands. Notice that the watershed was mostly unprotected all the way into Missoula. As usual, the state was in a subdivision boom, so MPC studied ways to develop its land. Scores of homes could be built along Rattlesnake Creek and beside the high lakes, a mix of recreational cabins and year-round residences. At the time, a gate blocked recreation access, but people climbed over to hike. Back then use was low and the unimproved parking lot could hold only six vehicles.

The Rattlesnake Mountains became a major focus of conservationists who feared the worst. Cass Chinske championed the idea of saving the whole place. He formed Friends of the Rattlesnake (FOR) and worked tirelessly with community leaders, NGOs, and the Five Valleys River Park Association to craft a way forward. Representative Pat Williams led the effort, introducing the legislation in Congress. Senators Max Baucus and John Melcher earnestly joined in. MPC was open to transferring its land to public hands, but it had

to be made financially whole for a deal to work. A defining goal and legislative momentum were needed.

In 1980, Congress passed the Rattlesnake National Recreation Area and Wilderness Act. It instructed the Forest Service to actively negotiate with MPC, but the power company had to agree with the appraised value of its holdings. Since the legislation did not include funding, a creative way of raising capital was required. In 1982, the price of MPC's 21,000 acres was $14.3 million. Big money—about $50 million in 2025 dollars. At first, various "surplus" Forest Service lands were offered to MPC, holdings that did not have significant public value. The idea was that the company would sell these stray properties to recoup its loss. This "land exchange pooling" approach had merit, but the right mix was never found. A deadline approached to get the deal done, 1984. That seemed far away, but time was running out.

Montana Power operated the Colstrip coal-fired plants in Eastern Montana. Even back then, coal was controversial. In those days, it was about strip-mining, community impacts, air quality, and prairie restoration. Few people talked about climate change. Negotiators grew frustrated by the lack of progress with a land-for-land approach, so another way had to be found. The solution seems simple in hindsight.

What did the power company need? Coal.

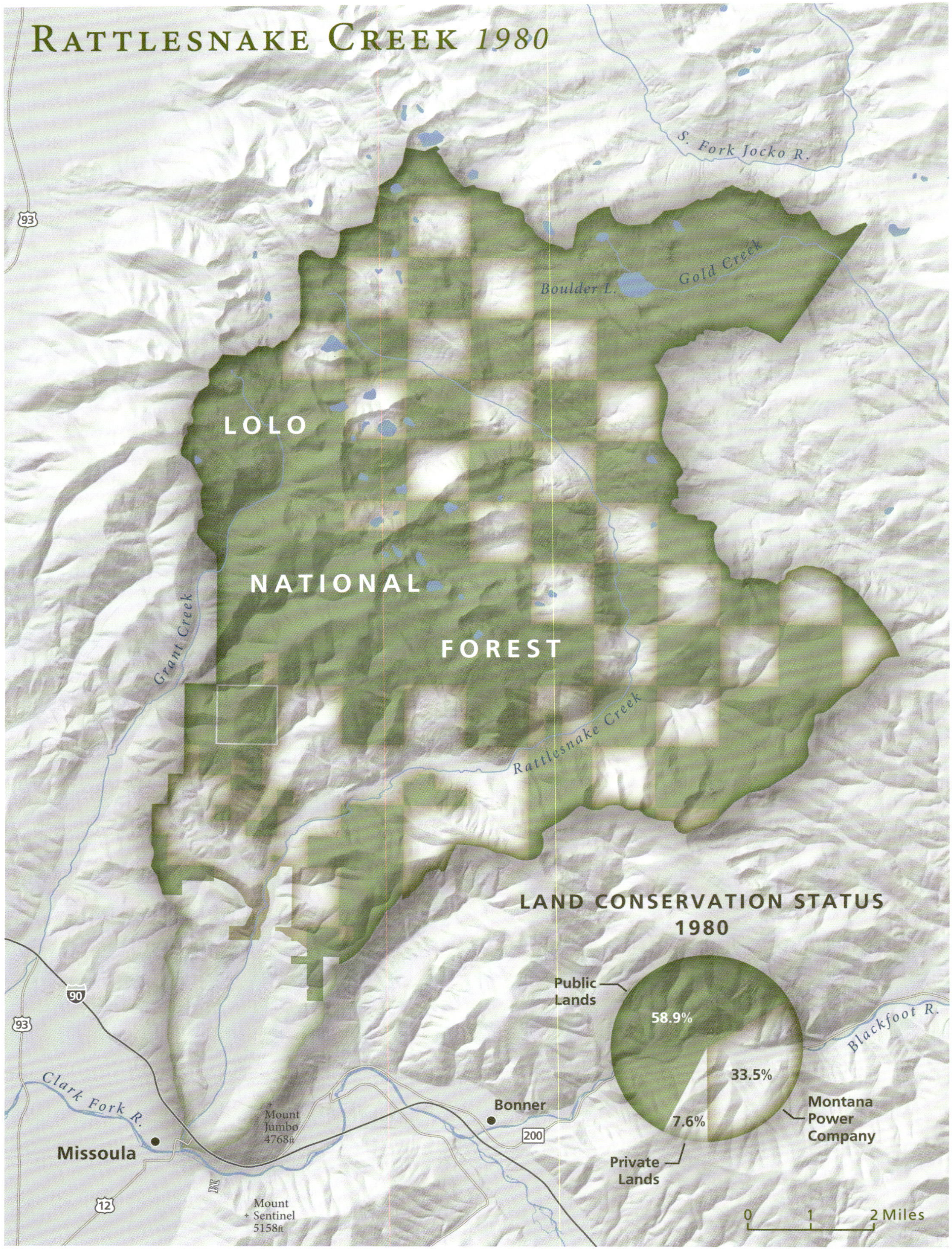

RATTLESNAKE CREEK 1980
S. Fork Jocko R.
Boulder L.
Gold Creek
LOLO
NATIONAL
FOREST
Grant Creek
Rattlesnake Creek
LAND CONSERVATION STATUS
1980
Public Lands
58.9%
33.5%
7.6%
Montana Power Company
Private Lands
Blackfoot R.
Clark Fork R.
Missoula
Mount Jumbo 4768ft
Bonner
200
90
93
12
Mount Sentinel 5158ft
0
1
2 Miles

RATTLESNAKE CREEK 2021
S. Fork Jocko R.
93
Boulder L.
Gold Creek
Rattlesnake Creek Trail
RATTLESNAKE
WILDERNESS
Grant Creek
RATTLESNAKE
NATIONAL
RECREATION
AREA
Rattlesnake Creek
LAND CONSERVATION STATUS
2021
Private Lands
4.8%
Public Lands
1.5%
92.9%
Conserved
Private
Lands
0.8%
Conserved
Public Lands
Rattlesnake
Greenway
90
93
Waterworks
Hill
City of
Missoula
Open Space
Blackfoot R.
Clark Fork R.
Bonner
Mount
Jumbo
4768ft
200
Missoula
Kim Williams Trail
12
Mount
Sentinel
5158ft
0
1
2 Miles

In Eastern Montana, most coal is owned by the federal government, specifically the Bureau of Land Management. An MPC subsidiary, Western Energy Company, was actively leasing federal coal managed by the BLM in projects cleared by the Environmental Impact Statement (EIS) process. Those public hearings were intense. What are the immediate and cumulative impacts of digging up coal and burning it? What are the alternatives? What about the Northern Cheyenne nation's status as a Class 1 Air Quality Area? Nearly half of all electricity used in Montana came from the Colstrip complex, a fact that even opponents had to acknowledge. Large amounts of coal were already available for extraction.

Mike Penfold of the BLM stepped forward with a bold idea. MPC would receive $14.3 million worth of "coal lease bidding credits" for mined coal based on the price it would bring in the marketplace up to the amount it was owed. This proposed transaction would require cooperation between the Department of Agriculture (Forest Service), Department of the Interior (BLM), MPC, the congressional delegation, the City of Missoula, and other parties. Such a deal had never been done. Conservationists were torn because they would split with their environmentalist friends. Souls were searched. Was this the right thing to do?

Pushing close to the deadline, MPC accepted the offer and the deal was completed. All 21,000 acres of MPC lands were deeded to the Lolo National Forest and then designated as part of either the Rattlesnake National Recreation Area or Wilderness. The distinction was based on the presence or absence of dirt roads, which explains the "stovepipe" of recreation area land along the creek. Today, backpackers bike up to the wilderness boundary and then hike into the backcountry. Most visitors stay within three miles or so of the main trailhead, sitting beside the water, picnicking, playing with their kids. We feel blessed to have this natural grace around us. Maybe we feel "greener" than those who don't revere the outdoors in the same way. But it's worth remembering how the land was protected.

This was essentially a coal-for-conservation trade. Critics argued that the coal could have been left in the ground *and* the Rattlesnake could have been saved, but that possibility never revealed itself. The truth is this: the coal involved had already been cleared for extraction under the National Environmental Policy Act (NEPA) process involving an EIS. It would have been burned anyway, and the Rattlesnake would have been lost. Trade-offs were considered, and a 61,000-acre marvel was saved for good. In the following years, 95 percent of the entire watershed was conserved in some way, up from 58 percent in 1970.

OPPOSITE: Rattlesnake Creek, autumn

Five Valleys Land Trust

In 1989, the river park association morphed into the Five Valleys Land Trust (FVLT). Conservationists decided it was time to join the national trend and form a local NGO to take on land-saving full time. County Commissioner Ann Mary Dussault strongly supported the idea since a land trust could move more quickly than government. Attorneys Helena Maclay and Bob Knight handled the incorporation process and much more. A rural planner named Amy Eaton moved over from the city to lead the group. She had a strong record in protecting land, so Amy turned her attention toward creating an open space corridor along Rattlesnake Creek. The land trust board was a who's who of Missoula conservation: Arnie Bolle was president of The Wilderness Society, Betty duPont was a retired rancher and conservationist, Chuck and Virginia Tribe worked for the Forest Service, and John Talbot and other business leaders brought their skills to the table.

The Montana Power Company owned 411 acres in the proposed corridor. Once again, it was receptive to selling, and once again the negotiations were daunting. The City of Missoula finally acquired an option to buy the land, which was then assumed by FVLT. In 1990, the trust received a $633,000 grant from the Land and Water Conservation Fund, and the next year the purchase happened. Senator Max Baucus was dedicated to the project, as was the rest of the congressional delegation. Amy Eaton and many others worked beyond the call of duty to make it a reality. This is a highly appropriate achievement given Missoula's Indigenous geography.

"Place of the Small Bull Trout" is what Salish people call the confluence of Rattlesnake Creek and the Clark Fork River. But for a century, an

MPC dam blocked bull trout from migrating very far up the creek. In 2021, the City of Missoula acquired the municipal water system including Rattlesnake water rights, the dam was removed, and riparian habitat was restored. Bull trout are now migrating up the creek all the way to the Rattlesnake Wilderness. If fish celebrate, they are. Imagine a bull trout finning her way farther upstream above a bed of gold, red, and green cobbles smoothed by thousands of years of flow. There are houses on the shore, but not enough to bother. A westslope cutthroat trout swims beside her, drawn to the same home water. She moves past the PEAS Farm, a community garden created by Josh Slotnik that feeds so many. A bit farther and the dam is gone, but the bull trout doesn't notice—Rattlesnake Creek is wild again. She swims onward, drawn by instinct. Somewhere in the recreation area she encounters other bull trout once trapped above the dam. Without knowing it, this migrant brings more native salmonid genes to the mix. She lays eggs to be fertilized, empowering evolution in a rewilding ecosystem. The creek flows back down to the river, so her return is fast. Connections lead everywhere.

Tracy Stone-Manning became director of the Five Valleys Land Trust when Amy Eaton moved on to a storied career with the Montana Land Reliance. Tracy is now the director of the Bureau of Land Management in Washington, DC. Five Valleys was fortunate to enjoy her leadership in completing several conservation easements in Pattee Canyon that enhanced a Forest Service recreation area. She also finished the Moon-Randolph Ranch easement in the North Hills. Bill Randolph was a hermit, so he had no peer pressure; he just loved the place. Today, it is used for agricultural education and can be reached by hiking trails linked to the city. Tracy's focus then shifted to the elephant in the room, Mount Jumbo. This major conservation project had been dreamed about for years, but now the threat escalated. Hundreds of homes could be built on its saddle and flanks. Millions of dollars were needed to buy the land and create an open space network. But how could that be done?

OPPOSITE: A walk in the woods

Outdoor writer and conservationist Greg Tollefson took the helm at Five Valleys after Tracy moved on to further success. Conservation is like a play where actors take their turn onstage. Greg knew his essential job was to conserve Mount Jumbo, a tremendous challenge. One day he and board member Chuck Tribe hiked to the top of the mountain to talk it over. Chuck simply said, "We don't have a choice. We have to do this."

In 1995, the Open Space Advisory Committee revealed the Missoula Urban Area Open Space Plan. County Commissioner Barbara Evans was stalwart in backing it. Mount Jumbo was singled out in Tollefson's words "for its unmistakable scenic value and its importance as critical winter range for elk and other wildlife." A hike to the summit takes you through waist-high bluebunch wheatgrass, blankets of yellow balsamroot, and stands of vanilla-scented pines. In early spring, rose-colored bitterroots sprout from tiny bulbs, a staple food for Native people and now an enduring sign of renewal. Black bears nap in hawthorn thickets and magpies swoop like frisky pterodactyls. A herd of elk leaves in warm times and returns in the winter when we cede the mountain back. Salish people call this place Sin Min Koos, which roughly translates as "Obstacle." That means terrain to go around, but Missoulians saw an opportunity.

Community interest in the project rose dramatically. John Klapwyk was the largest landowner on the mountain, with property appraised at $2.75 million. When all 1,600 acres of prime properties were tallied, the price tag rose to $3.3 million. Jumbo became the focus of an intense public campaign to pass a $5 million open space bond targeted at this project. Some 1,300 individuals, 350 businesses and community groups, and numerous foundations gave generously of their time and money. The Rocky Mountain Elk Foundation contributed $100,000 of "earnest money" to secure options to buy the land.

The bond was passed on November 7, 1995, a cold, snowy day. Voters overwhelmingly said "Yes!" to open space. The bond would cover $2 million of the cost of buying Jumbo, with Montana Fish, Wildlife and Parks and the Forest Service acting as partners in the deal. In 1996, $250,000 of Land and Water Conservation funding helped close the transactions. Greg Tollefson, the linchpin of the project, said, "This was a serendipitous junction of opportunity and collective interest." Finally, on March 2, 1997, all the deeds were transferred. The *Missoulian*'s headline read, "Jumbo Is Ours!" And it remains so.

Rock Creek

A perfect trout stream. An aquatic wonder flowing 54 miles from its source in the Anaconda Pintler Wilderness to its confluence with the Clark Fork River. The water is pure, the cobbles are clean, and the shady canyon keeps water temperatures ideal for trout. Salish and Pend d'Oreille people call the creek "Logs in the Water" because of its productive fishery. Natural logjams create nutrient-rich pools and back eddies where fish find refuge. Osprey and bald eagles watch from the trees, waiting for their chance. On a single day anglers can catch a "grand slam" of trout—rainbow, brown, cutthroat, brook, and bull. Moose dine in ponds and bighorn sheep loll in fields. You see full-curl rams on every trip. Talus slopes drape the base of mountains, pink quartzite blocks inscribed with black lichens. The Welcome Creek Wilderness (established in 1978) is entered by a footbridge over Rock Creek, with trails leading up to the windy crest of the Sapphire Mountains.

Rock Creek is only 30 minutes from Missoula, but the valley feels remote. The creek is bordered by Forest Service holdings, but private lands hopscotch along its course. Most of the place is ranched, but housing developments exist here and there. It's hard to blame people—a cabin on this creek is a family heirloom. The lower part of the watershed is in Missoula County, where land-use planning is rigorous, but the upper reach is in Granite County, where regulations are administered more casually. This was especially true back in the 1970s, when subdivision pressure first hit Rock Creek. Development is now separated by ag lands, Forest Service holdings, modest fishing lodges, and trailheads. The Rock Creek Road is called "dirt," but it's a potholed mishmash. Lousy access helps explain why the watershed hasn't been more heavily developed, but the real reason is a power line.

In 1985, the Montana Power Company and the Bonneville Power Administration (BPA) were granted a permit to build a massive 500-kilovolt power line that would cross Rock Creek five miles up the drainage. A lawsuit was filed to prevent it, headed by Montana Fish, Wildlife and Parks, Trout Unlimited, the National Wildlife Federation, and the Montana Wildlife Federation. According to Greg Tollefson, "They contended that the environmental assessment and planned mitigation measures were inadequate." Rock Creek was too valuable to lose. However, BPA had a legal right to condemn the right-of-way, so the power line would eventually be built despite objections.

Things took an unexpected turn. The lawsuit dragged on, and in 1986 Montana Power chose to settle by creating a $1.65 million trust fund for "Off-Site Mitigation" of ecological damage caused by the constructed power line. The Rock Creek Advisory Council was created, made up of agencies, NGOs, and landowners. Its mission statement was clear: "We are dedicated to protecting and enhancing Rock Creek's famous 'Blue Ribbon' trout waters, the health of its nationally acclaimed wildlife habits, its unusual biodiversity, and the open space beauty of the drainage." Salmonflies were studied, weeds controlled, stream sections restored, and an interpretive trail was built on the famous Valley of the Moon Ranch. However, the advisory council realized that housing development was the biggest threat, and conservation easements were the remedy.

By 1995, the Rock Creek Advisory Council had evolved into the Rock Creek Land Trust. All mitigation money was transferred by the state to this new NGO. Ellen Knight, a longtime Missoula conservationist, had worked for the council for years, and she saw the virtue of buying conservation easements. Ellen is a modest woman who recalls things this way: "Someone had to do it and I love Rock Creek, so it turned out to be me." Over the years, she built friendships with landowners up and down the creek. The landmark Handley Ranch was conserved. Lorraine and Jim Gillies placed an easement on their productive ranch west of Philipsburg. Petra De Groot and George Sinelnik protected their Diastole Ranch—it means the "resting phase of the heart." Tom Roy saved 160 acres up Stony Creek. In 1995, 1,400 acres of the Valley of the Moon Ranch were conserved by Roy and Susan O'Connor, and most of the land near the famous Ekstrom's Stage Station came

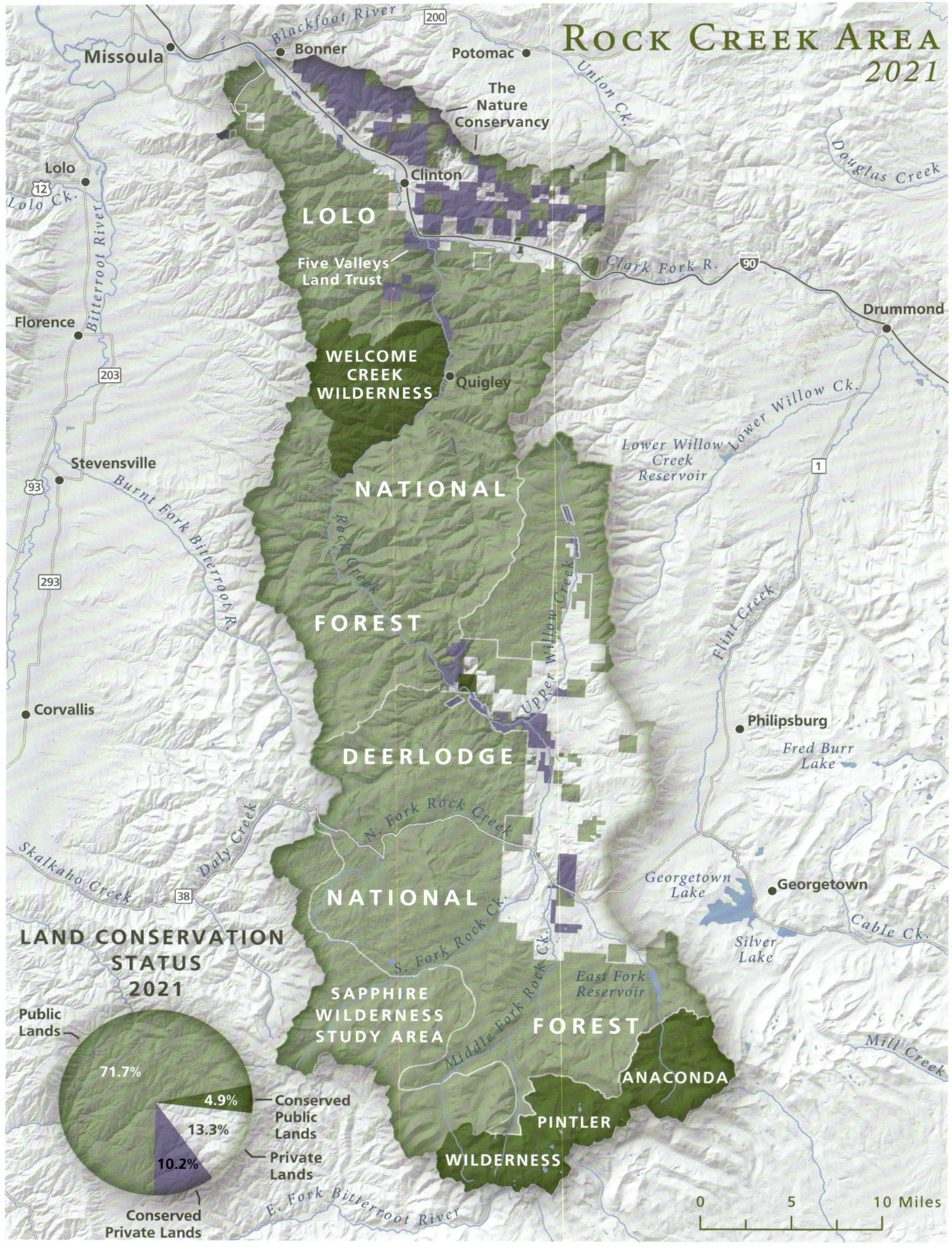
Rock Creek Area
2021
Missoula
Bonner
Blackfoot River
200
Potomac
Union Ck.
The Nature Conservancy
Douglas Creek
Lolo
12
Lolo Ck.
Clinton
LOLO
Five Valleys Land Trust
Clark Fork R.
90
Drummond
Florence
Bitterroot River
203
WELCOME CREEK WILDERNESS
Quigley
Lower Willow Ck.
Lower Willow Creek Reservoir
Stevensville
93
1
NATIONAL
Burnt Fork Bitterroot R.
Rock Creek
293
FOREST
Upper Willow Creek
Flint Creek
Corvallis
Philipsburg
Fred Burr Lake
DEERLODGE
N. Fork Rock Creek
Daly Creek
Skalkaho Creek
38
Georgetown Lake
Georgetown
NATIONAL
S. Fork Rock Ck.
Cable Ck.
Silver Lake
LAND CONSERVATION STATUS 2021
East Fork Reservoir
SAPPHIRE WILDERNESS STUDY AREA
Middle Fork Rock Ck.
FOREST
Public Lands
Mill Creek
ANACONDA
71.7%
4.9%
Conserved Public Lands
13.3%
PINTLER
Private Lands
10.2%
WILDERNESS
Conserved Private Lands
E. Fork Bitterroot River
0
5
10 Miles

under easement protection. The Marletto family eased another 2,300 acres in the upper watershed. The Clarks added 900 acres, and Bernie Nowak another 1,430. The Montana Land Reliance was a partner in getting some of these projects done. Over the years, Ellen Knight succeeded beyond measure, but here are the metrics.

Over 13,000 acres in the watershed are now protected by 24 easements, including more than 20 miles of creek frontage. The original $1.65 million trust fund has been leveraged to provide $30 million in on-the-ground conservation. Easements are held by the Five Valleys Land Trust, Montana Land Reliance, US Forest Service, and Montana Fish, Wildlife and Parks. The Rock Creek Trust is now a program of FVLT, its natural home.

In 2012, a 36-unit subdivision was proposed on the 300-acre Corra Ranch at the mouth of Rock Creek. Bulldozers and backhoes had already torn up the place, sending sediment into waterways. Such a development would destroy the scenic, ecological, and historical gateway to the drainage. Nick and Mary Babson and Don and Janemarie King bought the property to conserve it. They retained a small piece to place under easement and sold the rest to the Five Valleys Land Trust. Volunteers have restored the land, and accessible trails welcome all visitors. Five Valleys calls it the Confluence Property—a place where things come together.

In a way, trout found a way to save themselves in Rock Creek. The love of fish and the ecosystems that nurture them motivated people to honor a watershed. But to have trout you need cold, clear water and constellations of aquatic insects. Thanks to land conservation successes, Rock Creek still excels at both, and anglers conjugate the seasons by "the hatch." Early spring is the time of Skwala stoneflies, gray drakes, March browns, and blue-winged olives. Caddis flies emerge in May and continue for months. The salmonfly hatch in mid-June is a spasm of natural ecstasy. Millions of nymphs rise up from the streambed, cling to willows, mate, and then flump into the water where hungry trout feast. Salmonflies are two-inch-long, six-legged alms of protein—the foundation of the ecosystem and the local fishing economy. In peak summer, pale morning duns and yellow Sallies arrive, and by evening caddis "blitzes" drive trout loony. In August, spruce moths, Tricos, and grasshoppers mark the shift toward fall. In October, caddis and mahogany duns signal that snow is coming. As the days get colder, trout hunker down as they have always done and always will.

Conserving Rock Creek took more than good people and legendary trout; it took bugs.

The Ninemile Valley

In 1989, a female wolf shook her head as she woke up from being tranquilized. It was near Glacier Park and ranchers said she had killed livestock. Her sentence was relocation, but this black wolf escaped and made a run for it, searching for a home among her own kind. The radio-collar map of her wandering suggests desperation: the Great Bear Wilderness, a swim across Hungry Horse Reservoir, through Bigfork into the Mission Mountains, then the Upper Rattlesnake, over to the Bob Marshall Wilderness, back to the Flathead Reservation, and on to the Ninemile Valley. Intact conservation corridors made her journey possible. In the Ninemile, she finally met another wolf, a scruffy gray male that biologists didn't know was there. The wolves mated, occupied an old coyote den, and emerged in the spring of 1990 with six pups.

"Holy mackerel, what are they doing in the Ninemile?" Mike Jimenez remembers asking. Mike is a wolf biologist who worked for the US Fish and Wildlife Service for more than 30 years. At that time, he thought there were no wolves south of Glacier. The den was on Bruce and Ralph Thisted's historical ranch, so they were the first locals to spot the animals. "It was a year or so before others saw them," Ralph said. Soon, the female wolf was killed and the male was left to raise the family. The Thisteds took home movies of the pups chasing grasshoppers and playing in their meadow. Then the male was killed and "the pups were orphaned," Ralph said. Mike Jimenez's job was to keep them alive. One day, he howled like a father wolf and the young ones ran out excitedly to greet him. It surprised everyone. Mike carried road-killed deer to the pups and that winter they made their first kill. The future of wolves seemed brighter, but it remains uncertain.

Wolf

The Ninemile Valley is rich in big game, clean water, shady forests, and lairs for wolves to hide from trouble. They have their reasons. Over the decades, about 700,000 wolves have been killed in Montana—trapped, poisoned, or shot. In 1995, wolves were reintroduced to Yellowstone National Park because even that vast wilderness had been swept clear. However, the Ninemile is ranch country, and the arrival of this apex predator was and is controversial. Starry and Cat-Man, two prized llamas, were attacked and eaten. The Thisteds lost a dog to the very wolves they admired. Calves and sheep were killed, reducing ranch incomes. In turn, wolves got shot, were hit by vehicles, or vanished.

However, many local landowners supported wolves or were willing to coexist with them. Ralph Thisted said, "Over the decades, we've lost hundreds of sheep and cows to bears, coyotes, and mountain lions. It's tough, but losses are just part of raising livestock." His wife, Bette, agrees. "There's just got to be room enough here for everyone and everything."

The Thisteds lived up to that. Getting older, they decided to sell the ranch to a buyer who would protect it with a conservation easement. At that time, Andie MacDowell was searching for a quiet place in Montana to raise her family. The actress had deep Scots Irish roots in Appalachia, and as a kid, she "escaped into the forest for peace." Andie visited the Thisted ranch and the brothers led her to a grove of western red cedars. "I couldn't speak," she said. "So much diversity and beauty—it looked like fairies lived there." Andie remembers thinking, "I'm in trouble, I really love this place, but I'm not rich." She was starting out in the movie business, and buying the ranch was a real financial leap, but she took it. "I had an open heart, but Phil Tawney and Grant Parker guided me in doing the right thing by donating a conservation easement to the Montana Land Reliance." Andie speaks plainly about it. "Me and my three kids are very proud of that easement. We love the land with all our hearts."

So do her neighbors. The Hagers placed an easement on their nearby ranch. Bette Thisted

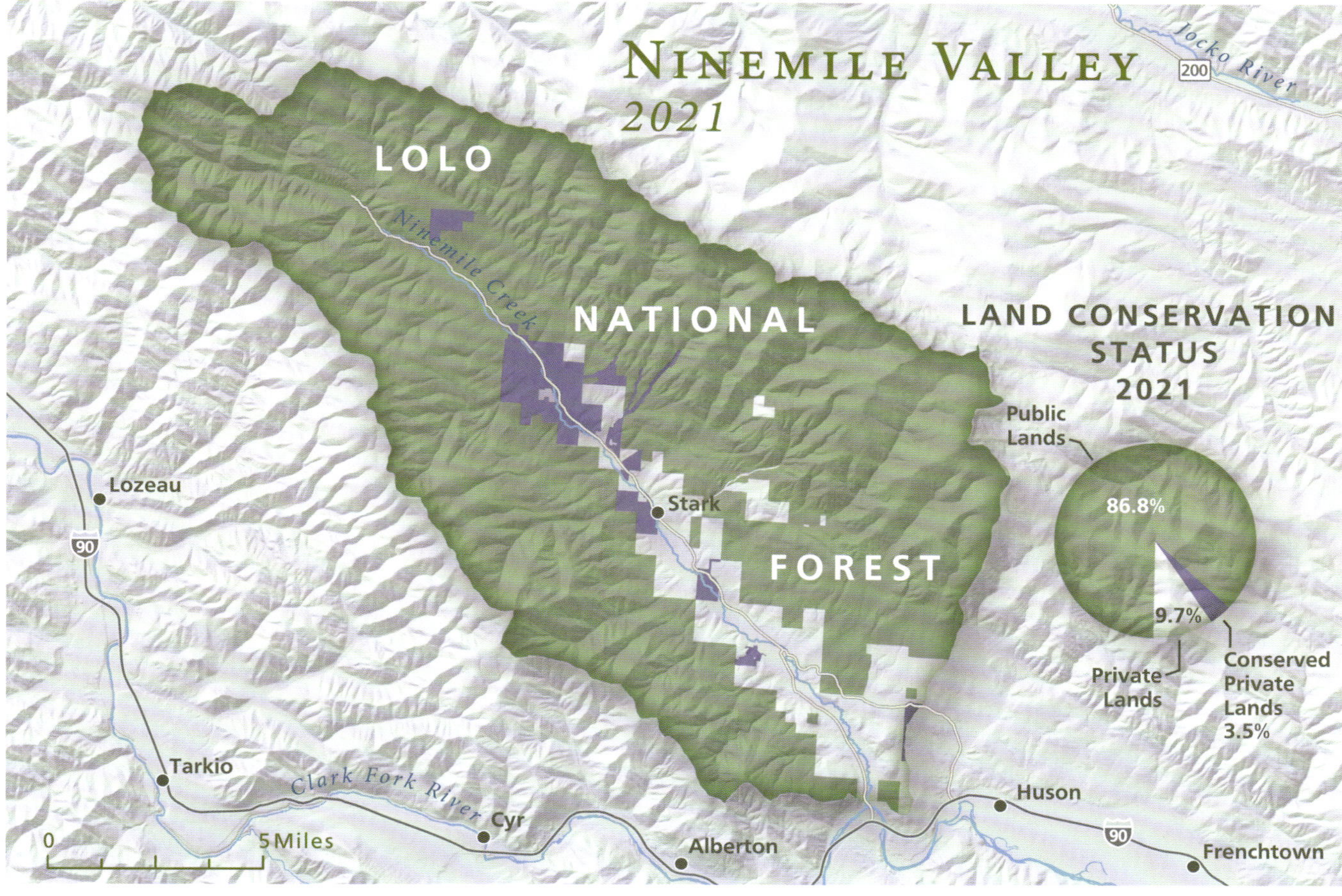

donated an easement on a half section of forest habitat. Another 2,000-acre conserved property extends across the valley. The Montana Land Reliance holds easements on over 4,000 acres in the watershed, with the Rocky Mountain Elk Foundation and the Five Valleys Land Trust safeguarding 1,000 more. Easements protect about 25 percent of all private land, and 90 percent of the total watershed is conserved. Andie MacDowell quietly led by example, but the Thisteds and other ranchers were land stewards who made it possible. As were the Salish and other Indian nations before them who lived in this valley below Ch-paa-qn Peak. The Salish name means "Shining Mountain."

Today, the wolf population hovers around 10—one pack out of dozens in Montana. Some say it's five times that, but the Ninemile easements are about more than wolves. They display a working compromise between ranching, open space, and species of all kinds. Challenges remain. The human population grows as subdivisions fill the lower valley. Ranches, public land, and houses intermingle, so wolves must navigate a land-use minefield. Mike Jimenez said, "There are really nice people here and some have very strong opinions about wolves, but everybody is friendly and gracious. They'll say, 'Don't kill all the wolves, just manage them more closely.'" Mike believes that wolves who become acclimated to killing livestock must be shot. He sees that as a tough but necessary step in maintaining cultural, political, and ecological balance. Ranchers are also paid for animals killed by wolves and other predators by the Montana Livestock Loss Board. It's a fair system. Mike adds, "Wolves are resilient. As long as we let them, they bounce back."

In the Ninemile Valley, they have a chance.

The Alberton Gorge

Whitewater rivers are where we surrender control. Paddling means adapting to what is in front of us, not forcing our will. The Clark Fork River is unaware of our schedules and fears, but it means no harm or good. It just surrenders to gravity, cutting into bedrock, rolling boulders into rapids.

The forested canyon walls of the Alberton Gorge provide habitat for elk, deer, and bears. Falcons build aeries on purple quartzite cliffs and the fishery is just fine. Greg Tollefson wrote that the gorge "guards a vital undeveloped link for wildlife to move back and forth from the Northern Continental Divide ecosystem to wild lands in the Selway-Bitterroot and beyond."

This eight-mile stretch of whitewater attracts 30,000 paddlers and floaters every year. You put in at Cyr Bridge and take out at Forest Grove, with lots of waves and splashes in between. There are 15 named rapids, gentle at first, until you encounter Tumbleweed, Boat Eater, and Fang. These class 3 rapids flip the boats of unprepared or unlucky souls. Survive that and you stop for lunch at Fish Creek, walking to an aquamarine pool. It's both a relief and a disappointment when the whitewater is done. You're cold and soaked, but you pledge to run the gorge again. Thanks to a conservation deal with the Montana Power Company, you can.

In the early 1990s, MPC held all the cards, owning land along the gorge and the right to build a dam on the Clark Fork River. Dale Harris, founder of the Great Burn Conservation Alliance, stressed how vital it was to protect the gorge. Government agencies, recreationists, and finally MPC itself contacted the Five Valleys Land Trust to make that happen. The idea was to buy all power company lands and transfer them to management by Montana Fish, Wildlife and Parks. The right to build a dam would also be relinquished, a low-probability outcome, but worth eliminating in this special place. In 1998, the River Network, a national group, finalized a purchase agreement for power company lands. But in 2003 NorthWestern Energy bought MPC, the River Network left the project, and the Five Valleys Land Trust took over. A local attorney and paddler named Peter Dayton agreed to handle all legal work. No one knew how much maneuvering would be required.

Here's a sense of the complexities. Five Valleys secured a $750,000 line of credit and then oversaw the exchange of 70 parcels between the Forest Service, Montana FWP, the Montana Department of Natural Resources and Conservation, and NorthWestern Energy. An Environmental Impact Statement was prepared, including numerous

public hearings. The original purchase agreement was extended more than 10 times because of delays. Greg Tollefson earned deep battle scars, but he got the project done.

In 2004, all power company lands were transferred to Montana FWP. The gorge was safe—no houses would be built, no dam would be constructed, and river access sites were provided for boaters. This whitewater canyon is now a key part of the region's identity, a place to experience cold-water wildness and then return to the warmth of the sun.

Barry Lopez wrote, "To put your hands in a river is to feel the chords that bind the earth together." He must have run the Alberton Gorge.

Many Mountains to Climb

In 1998, Wendy Ninteman entered the story as Greg Tollefson moved on to other creative pursuits. She brought a strong environmental consulting background to her role as executive director of Five Valleys. Under Wendy's leadership, the land trust expanded its strategic planning and attained accreditation from the Land Trust Alliance, a national umbrella group based in Washington, DC. One of her first projects happened when a lifelong rancher named Jim Cusker announced he wanted to place an easement on his family's land in Grass Valley along the Clark Fork. Jim held Five Valleys in high esteem and would go on to serve with distinction as a board member for 25 years. He inspired his neighbors, the Vallejo family, to conserve their farm. Likewise with Joe Boyer, who placed an easement on 1,000 acres of prime farmland and riparian wildlife habitat. Peregrine falcons, golden eagles, and 80 other bird species benefit from these Grass Valley projects. Joe Boyer said, "My father always told me just take good care of this place while you're here. Pass it on the same way. I'm sure when I signed the papers with Five Valleys, all my ancestors were smiling someplace."

Mayor Mike Kadas was a strong ally of the Five Valleys Land Trust. Mike worked closely with Wendy Ninteman to complete the protection of Mount Sentinel by purchasing 475 acres from the Cox family. City Council member Chris Gingerelli represented that district and played a crucial role. Mayor Kadas said, "When we see these hills, coming in by air or highway, or looking up from our morning coffee, we remember who we are. As a city, we stop at this boundary." Wendy Ninteman sighed with relief and simply said, "This property is one of Missoula's most spectacular open space gems." But there was another mountain to save—Dean Stone.

Time for another open space bond—this time for $10 million. In 2007, true to their nature, Missoulians passed it with over 70 percent of the vote. Pelah Hoyt from the Jocko Valley took the lead in generating support. She went on to complete 15 years with Five Valleys, including negotiating purchases on Mount Dean Stone. Pelah also helped improve connections with the Confederated Salish and Kootenai Tribes.

Dean Stone is one of Missoula's key peaks along with Jumbo and Sentinel. The Barmeyer family owned much of the land and generously agreed to protect it. Allen and Candace Fetscher, wildlife biologist Jack Lyon, and other landowners agreed to direct or bargain-sale purchases of much of the rest. Open space now connects the city with public lands and a 1,000-acre community forest purchased from The Nature Conservancy by Five Valleys. The High, Wide, and Handsome Trail winds through the Mount Dean Stone Preserve and connects with a corridor leading to Yellowstone. Its namesake was Arthur Stone, the founder and dean of the School of Journalism at the University of Montana. You can see the peak from his old office, so it's certain he would be proud. Inspired by the project, three more families stepped up to complete easements at the base of the mountain—the Lines, Rimels, and Haydens. Next time you run into them, say thanks.

Meanwhile, Wendy Ninteman and Five Valleys continued to complete easements in the Blackfoot, the Bitterroot, Mineral County, and beyond. Through the years leadership changed—Kate Supplee, Jim Berkey, Whitney Schwab—but forward motion remains strong. The Five Valleys Land Trust has conserved over 135,000 acres. The Montana Land Reliance, The Nature Conservancy, the Trust for Public Land, The Conservation Fund, cities, counties, and land management agencies have added thousands more. But, as Carl Safina wrote, "Facts alone don't save the world, hearts do."

An Intentional Community

There are places like the Missoula Region where land conservation is the foundation of a good life. As you've seen, this is also true for corporations like NorthWestern Energy, formerly the Montana Power Company. Open space is called Green Infrastructure for the ecological and public health benefits it provides. It also grants us emotional solace from our troubles and the freedom to express our joy. Conserved land becomes a beloved family member, expanding our sense of place. Newcomers often pick the Missoula area because of the abundant open land, something geographers call the "intentional community effect." People choose a landscape that matches their soul while bringing talents and resources with them. Critics argue that open space causes home prices to rise because the land can never be built on. Perhaps, but developing the landscape willy-nilly is economically unsound, and finding the sensible center seems wise.

In the Missoula Region, the land has standing. This means that Nature has a place at the negotiating table—and we're listening.

FURTHER READING

Five Valleys Land Trust. www.fvlt.org (this expansive website contains detailed content about specific conservation projects and land purchases).

"From Ridge to River: Conserving Open Space in Missoula, Montana." 2002. Jed D. Little. Master's thesis, University of Montana.

"A Gracious Civic Enterprise: Five Valleys Land Trust History, 1972–2022." Greg Tollefson. 2022 (excerpted on the Five Valleys Land Trust website).

Grizzly Hackle. www.grizzlyhackle.com (the explanation of trout and insects on Rock Creek is exceptional).

"Ninemile Wolf Stories." 1992. Sherry Devlin. Timber Wolf Information Network. www.timberwolfinformation.org (Sherry's fine account is the source of some of the quotes and details in this chapter).

The Ninemile Wolves. 1992 (with more recent printings). Rick Bass. Houghton Mifflin.

CHAPTER 13

Northwest Montana

This is a landscape of productive timberlands, binational rivers, and rumors of reindeer. The Flathead and Kootenai watersheds begin in the Rocky Mountains of the United States and Canada, with the Kootenai migrating back and forth across the border. The Northwest is the wettest region in Montana, with forests reminiscent of the Cascades, where western red cedar, hemlocks, grand fir, and larch grow beside more typical species like ponderosa pine and Douglas fir. In the late nineteenth century, the federal government allocated railroad land grants in alternating sections along the route of the Northern Pacific to pay for construction and fill the landscape with settlers. When homesteading largely failed in Northwest Montana, timber companies accumulated vast properties adjacent to Forest Service lands. Today, most old growth and second growth has been harvested, but significant habitat remains for elk, moose, deer, and an array of bird species. The landscape is well watered with plentiful lakes, including Flathead—the biggest freshwater body west of the Great Lakes.

However, the region has a split personality. The Kalispell-Whitefish area is being rapidly subdivided and developed because of its proximity to Glacier National Park and the Whitefish Mountain Resort. Flathead County has 112,000 people. West of Lake Koocanusa is a remote part of "old Montana" where logging, hunting, and fishing are mainstays of life in towns like Libby and Troy. Lincoln County has only 22,000 people—five times smaller. Sanders County has 13,000. The varying approaches to land conservation reflect those clear geographic differences, but across those boundaries is a unifying fact: this is one of Montana's prime homelands for grizzlies. Land management adapts to that knowledge, but there is a surprise out in the trees. Woodland caribou (reindeer) are still occasionally seen (even by adults) in the thick woods around Yaak. While caribou hunting is no longer permitted, other big game species thrive. Montana Fish, Wildlife and Parks has classified the "Heart of the Salish Priority Area" as a major big game winter range and migration corridor across the region. Kootenai place-names reveal the same ecological and hunting importance. Yaak means "Arrow," and the tribe says the course of the Kootenai River bends into the shape of a bow. In both the rural Northwest and the growing Flathead Valley, land conservation is expanding.

Kootenai Falls

Shrines are often made of rock—Bears Ears, the Vatican, the Western Wall, Mecca, Notre-Dame Cathedral, Sri Pada, Mount Kailash, and Uluru. Kootenai Falls is created by water cutting into billion-year-old quartzite and limestone along a one-mile reach of river. The flow thunders over ledges, cascades through chutes, and surges through side channels where torrents of white-water end in swirling turquoise pools. Erosion continues to create this sacred site of the Kootenai people. Waterfalls are ancient and forever young.

In the early 1800s, Canadian explorer David Thompson used the Kootenai River as a navigational guide. His party portaged around the falls on a steep trail following Indian cairns. Thompson was fearful. "The least slip would have been inevitable destruction . . . in once falling, no stop until precipitated into the river." But for Kootenai people, the falls was and remains a place to receive counsel from the Earth. Visions are sought and the power of the place brings humility and reverence, but this shrine was nearly lost.

In 1978, a utility consortium proposed building a hydroelectric dam at Kootenai Falls. A reservoir would inundate the cataract, destroying its spiritual and ecological essence. The Federal Energy Regulatory Commission (FERC) rarely denied applications, so the situation was dire. Six years

Kootenai River

earlier, Libby Dam had been built upstream on the Kootenai River, backing up Lake Koocanusa all the way into Canada. Native people had seen dams before. Lawrence Kenmille (Kootenai) said, "For many years we kept the importance of the falls to ourselves, but with the dam proposal, we have to stand up to protect it. Kootenai Falls has been an aboriginal area and it has religious significance to us. We went there for a thousand years to pray. We do not build our own monument. Nature builds a monument for us." Attorneys Walter Echo-Hawk (Pawnee) and Steve Moore, along with conservation groups, petitioned FERC to deny the dam license because it would cause ecological harm and severely damage the religious rights of Kootenai people. The case dragged on for years. In 1984, after a 13-week trial, the judge denied the permit based largely on spiritual grounds. Three years later, all appeals were exhausted and Kootenai Falls was saved.

Today, visitors cross a swinging bridge over the falls, feeling its power. Sitting on the bank gives a different perspective. We are not above this place; we defer to it. Non-Native people spend a few hours here and leave with appreciation. Six hundred generations of Native Americans have seen this as holy water.

There are times when land conservation requires more than negotiation and voluntary methods. Some places are priceless and irreplaceable. Kootenai Falls is one of those.

Thompson Chain of Lakes State Park

A corridor of 18 lakes nearly forms a circle in the upper reaches of the Thompson River. Loon, Lilypad, Horseshoe, Crystal, Upper and Lower Thompson, McGregor, and more. This 20-mile fishery is brimming with trout, northern pike, yellow perch, largemouth bass, and kokanee salmon. Bald eagles dine on fish, otters scurry in the willows, and turtles sunbathe on logs and slip into the water like silent stones. Lakes warm up in the summer, making boating and swimming a delight. This country is highly coveted for recreational cabins and commercial development.

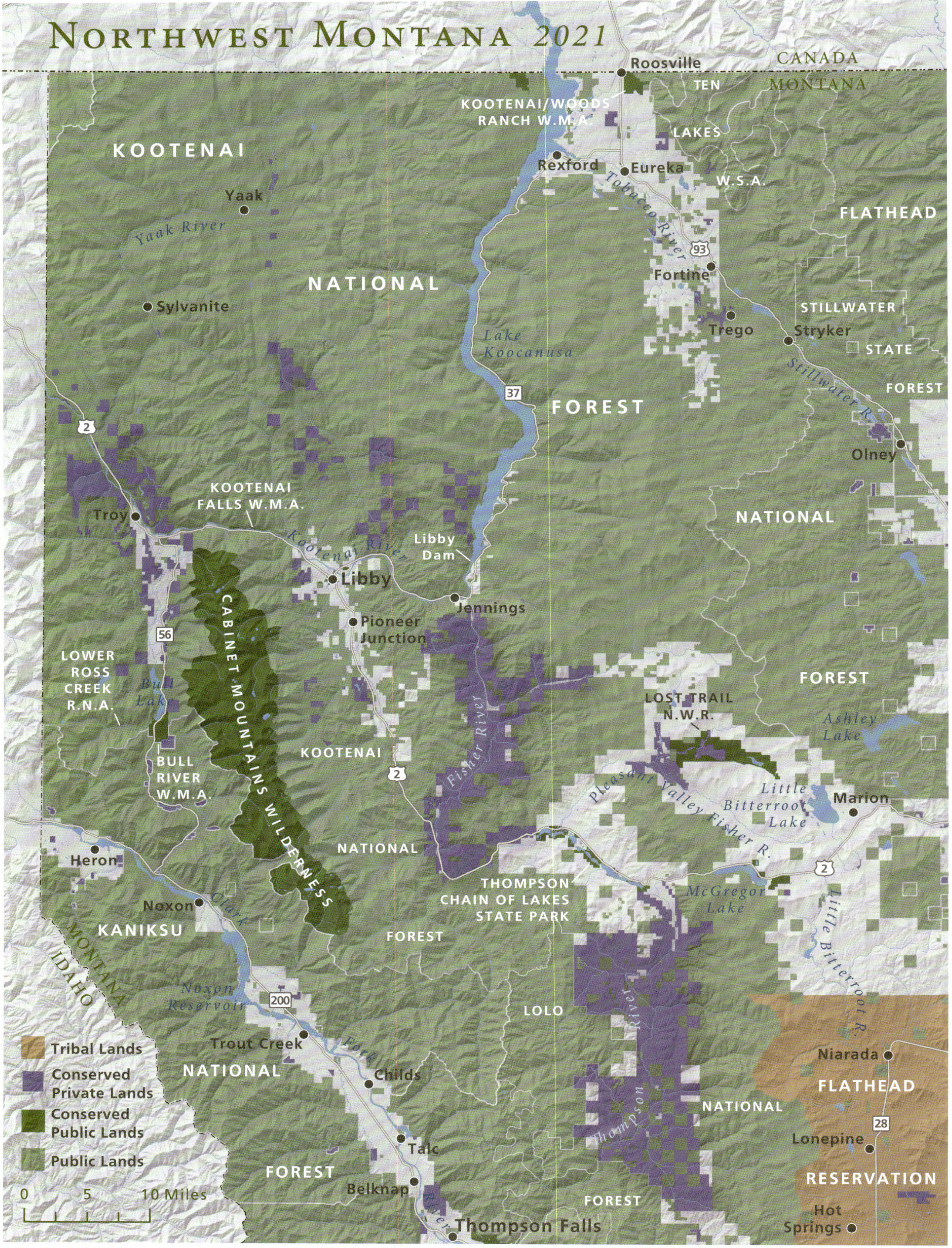
NORTHWEST MONTANA 2021
CANADA
MONTANA
Roosville
TEN
LAKES
W.S.A.
KOOTENAI/WOODS RANCH W.M.A.
Rexford
Eureka
Tobacco River
KOOTENAI
NATIONAL
FOREST
Yaak
Yaak River
Sylvanite
FLATHEAD
Fortine
Trego
Stryker
STILLWATER
STATE
FOREST
Stillwater R.
Olney
Lake Koocanusa
NATIONAL
FOREST
Troy
KOOTENAI FALLS W.M.A.
Kootenai River
Libby Dam
Libby
Jennings
Pioneer Junction
CABINET MOUNTAINS WILDERNESS
LOWER ROSS CREEK R.N.A.
Bull Lake
BULL RIVER W.M.A.
KOOTENAI
NATIONAL
FOREST
Fisher River
LOST TRAIL N.W.R.
Ashley Lake
Pleasant Valley Fisher R.
Little Bitterroot Lake
Marion
McGregor Lake
THOMPSON CHAIN OF LAKES STATE PARK
Little Bitterroot R.
Heron
Noxon
Clark
KANIKSU
IDAHO
MONTANA
Noxon Reservoir
Trout Creek
Fork
Childs
NATIONAL
FOREST
Talc
Belknap
River
Thompson Falls
LOLO
Thompson River
NATIONAL
FOREST
Niarada
FLATHEAD
Lonepine
RESERVATION
Hot Springs
Tribal Lands
Conserved Private Lands
Conserved Public Lands
Public Lands
0 5 10 Miles
2
37
93
56
200
28

Thompson Chain of Lakes

Resorts, retreats, and homes are built, with more development coming.

In 1999 the Richard King Mellon Foundation bought 4,000 acres and donated the land to Montana Fish, Wildlife and Parks. It's now called the Thompson Chain of Lakes State Park—a public resource we can enjoy forever. Boating, fishing, camping, dunking on a hot day, viewing wildlife; all pleasures remain intact. Richard King Mellon was a successful Pittsburgh banker. Some complain about rich Easterners influencing Montana, but he deserves respect for saving a natural treasure. Mellon served in the Army during World War I and World War II, receiving the Distinguished Service Medal. Next time you land a fish here, thank him for all his service.

Wood ducks

Lost Trail National Wildlife Refuge

Pleasant Valley is north of the Chain of Lakes, but ecologically connected. The landscape is a mosaic of lakes, wetlands, riparian corridors, bunchgrass prairies, and coniferous forests. For over a century, ranchers impounded water, dug drainage canals, channelized creeks, and irrigated fields. Natural or not, all that water attracted wildlife in incredible abundance. In 2000, the Lost Trail National Wildlife Refuge was created when the Montana Power Company donated 3,112 acres in Pleasant Valley. Another 4,773 acres was sold to the US Fish and Wildlife Service to mitigate the impacts of Kerr Dam on Flathead Lake. Federally owned wetlands along the north shore of the lake were being lost because of fluctuations in water levels caused by dam operations. In 2015 the Confederated Salish and Kootenai Tribes bought the dam and renamed it Séliš Ksanka Ql'ispé. Their intention was to better control lake levels while still earning income for tribal programs.

The Lost Trail National Wildlife Refuge is where seasons stir wildlife cycles. In spring, owls and Canada geese nest, and then green-winged teal and other waterbirds arrive at Dahl Lake. In summer, the water is surrounded by meadows where elk and moose calves kick up their heels. In fall, aspen and larch turn gold and creatures migrate away or prepare for the cold. Winter is a bottleneck for big game, who find refuge in south-facing grasslands. Nature has no off-season, and it requires lots of room to thrive. In the following decades, timber companies, conservationists, and agencies provided it.

Working Timberlands

Logging built the United States *and* created deep conflicts about conservation. Clear-cutting old-growth forests and its negative effects led to the creation of the Multiple-Use Sustained-Yield Act in 1960. Impacts of timber harvesting on water quality, wildlife, and communities now had to be considered in national forest plans. The idea was to use "sustainable" practices, which meant "cut the amount of timber each year that can be cut every year forever." And do so using ecologically sound methods.

However, national forest management didn't change much until publication of the *Bolle Report* in the *Congressional Record* in 1970. Arnie Bolle, dean of the Forestry School at the University of Montana, led an investigation into clear-cutting, terracing, shoddy roadbuilding, habitat loss, and stream siltation in the national forests south of Missoula. The report began bluntly: "Multiple use management, in fact, does not exist as the governing principle of the Bitterroot National Forest." It went on to say that "their overriding concern is timber production," with no respect for the Multiple-Use Sustained-Yield Act. Senator Lee Metcalf was from nearby Stevensville and strongly supported those findings. However, the Bitterroot Forest wasn't alone; that same mindset existed across the system. Over the years, a combination of lawsuits, economics, and a shift in agency culture led to meaningful improvements in management. Timber harvests on private land were not impacted by the *Bolle Report*, so trees were cut with little concern for the future.

Great gray owl

In the 1990s, the future arrived in the heavily logged forest of Northwest Montana. Champion International and Plum Creek Timber Company realized that most big trees had been cut and profit margins were falling. The companies decided to sell much of their land to Stimson Lumber Company, which was committed to the Sustainable Forestry Initiative's goals as an exercise in good stewardship and good business. These goals include

- Sustainable Forestry
- Forest Productivity and Health
- Protection of Water Resources
- Protection of Biological Diversity
- Aesthetics and Recreation
- Legal Compliance
- Respect for Indigenous Rights

The Sustainable Forestry Initiative has certified 400 million acres in the United States and Canada, including land owned by Stimson. The Trust for Public Land (TPL) saw an opportunity, and negotiations led to Stimson selling conservation easements on over 50,000 acres of its working lands. The company made a long-term commitment to the woods, with metrics and monitoring to ensure it. Public access is allowed for hunting, fishing, and wandering as long as people don't abuse the privilege. Montana Fish, Wildlife and Parks holds the easements and manages recreation.

TPL also works extensively in Northwest Montana to conserve timberlands and connect conserved places. Its Montana Great Outdoors Project is now a "priority" of the US Forest Service, and the NGO was awarded $20 million by the Forest Legacy Program. Forestland easements were the goal.

In 2020, Weyerhaeuser Corporation announced it was selling all 630,000 acres of its timberland in the state. A company called Southern Pines Plantations (SPP) bought Weyerhaeuser's land in Northwest Montana. The company buys and sells productive lands, often working with agencies and conservation NGOs. TPL partnered with the USFWS to purchase conservation easements from the company on 38,052 acres within the Lost Trail Conservation Area—ecosystems surrounding the national wildlife refuge. The land is now sustainably managed for timber production, and public access is allowed. The property stays on the tax rolls and supports sawmills. TPL's Northern Rockies director Dick Dolan said, "This project wouldn't have been possible if not for the vision of Southern Pines Plantations and the USFWS. This helps keep Montana, Montana—and protects access to our favorite hunting spots, hiking trails, and secret fishing holes. This project represents all the best aspects of conservation in Montana."

A company called Green Diamond agreed. In 2021, it sold a 142,000-acre conservation easement on its working timberlands nearby. TPL helped structure this immense easement, which is now held by Montana Fish, Wildlife and Parks. The forests are also managed using Sustainable Forest Initiative standards, and a Habitat Management Plan guides timber operations. TPL's Great Outdoors project is now a clear success.

The US Forest Service established the Forest Legacy Program to broaden conservation efforts, and it succeeded. By 2021, this effort had permanently conserved 261,000 acres, "including 243,000 acres of conservation easements and 18,000 acres of fee title acquisitions." The money comes from the Land and Water Conservation Fund, which is generated by federal offshore oil and gas revenues.

More than 200,000 acres of Northwest Montana forests have now been conserved. In time, more private timberlands will be protected by these sensible agreements. This entire biodiverse landscape will remain mostly intact, since 73 percent of Lincoln County and 52 percent of Sanders County lie within the Kootenai and Kaniksu National Forests. The Montana Land Reliance, the Flathead Land Trust, and other groups also hold easements across the region, linking large parcels with constellations of smaller ones. As mutual interests align and compensation is provided, relationships lead to results. Even the old divides between logging, wildlife, and recreation are becoming connections.

Vital Ground

You know a grizzly and his name is Bart the Bear. This 10-foot-tall, 1,500-pound "actor" appeared in 13 movies, including *The Clan of the Cave Bear*, *Legends of the Fall*, *The Bear*, and *White Fang*. Doug and Lynne Seus adopted Bart as an orphan out of their profound respect for the species. In 1990, they cofounded a grizzly bear conservation group called Vital Ground that works in Northwest Montana, the Greater Yellowstone, and wherever the immense bear lives. Douglas Chadwick is another cofounder, a wildlife biologist who scientifically grounds the group. Doug got his master's degree at the University of Montana in 1974 as one of a distinguished crop of conservationists emerging from that campus. He studied mountain goats in college, and then wolverines, before moving on to his lifelong passion—bears. Doug works closely with the National Geographic Society in producing books and documentaries designed to build public support for conservation. His book *Four-Fifths a Grizzly* reveals our close genetic kinship with the species—80 percent of our DNA is the same. At the Vital Ground Foundation, Chadwick helps focus land purchases and easements on prime habitats and migration corridors for grizzlies. The group's mission is clear: "To protect and restore North America's grizzly bear populations for future generations by conserving wildlife habitat and by supporting programs that reduce conflicts between bears and humans."

After 33 years, Doug Chadwick has a deep perspective. "Vital Ground is a grizzly bear-centric group, but they're an indicator species at the landscape level, much like wolverines and bison. Where do they live, where are they moving, and how can those large habitats be conserved? One time we gathered 60 biologists in a room and had them stick pins on a map where grizzlies showed us they needed a hand. Combined, that's centuries of on-the-ground knowledge. Vital Ground relies on science."

Alaska is home to 95 percent of all grizzlies on Earth, more than 30,000 animals. Doug Chadwick reports that Montana has about 2,000, a 96 percent decline from the time of Lewis and Clark. Habitat loss, bounties, and predator control took a brutal toll. In response, Vital Ground has conserved or enhanced over 630,000 acres of habitat to aid in griz recovery, with meaningful portions of that in Big Sky Country. Its "One Landscape Initiative" works to protect 188,000 additional acres in 33 key landscape linkage areas in the Northern Rockies and Great Plains.

Vital Ground has built scores of partnerships to get that done, especially with Montana FWP. Chadwick sees the role of Vital Ground clearly. "We can't lobby and don't hold placards in the street. We partner with others and are apolitical in order to get projects done. It's an optimistic way of doing business. So many people concerned with habitat are despondent, but to me, voluntary land conservation gives you tangible hope. It's a way of breaking down the divide between what's good for people and what's good for nature. Conserved land changes our physiology for the better. Our heart rate slows, our blood pressure drops, and our immune system is strengthened. If you want people to be happier, healthier, and live longer lives, then conserve land. What land trusts do also improves our communities. Whenever someone complains about land being 'locked up' I ask two simple questions. Is Montana getting poorer and are fewer people coming here because we've conserved lots of land? Of course not. Montana still looks like Montana because we've kept so much of it in good shape."

Doug returns his focus to grizzly bears. "They're an umbrella species—where bear habitat exists, all other species thrive. Grizzlies are opportunistic omnivores not carnivores, so there is little reason for fear." They eat mostly plants, berries, and bugs, but they will consume dead animals or kill weak ones, recycling nutrients and fertilizing the soil. Grizzlies are a key part of food webs, and the land is much healthier with them present. The Cabinet-Yaak Ecosystem supports only 50 grizzlies split into two populations. Half of them live within the 94,000-acre Cabinet Mountains Wilderness, formed in 1964. Chadwick says, "Corridors are emerging between places like the Cabinet-Yaak region and the northern Bitterroots, so that's the kind of place we need to work."

In Northwest Montana, Vital Ground owns four properties and holds easements on 2,500 acres. The Hubbard Farm easement conserves 1,040 acres of grizzly bear habitat along the Kootenai River, offering safe passage across the valley. True to Vital Ground's style, it was a cooperative project with the US Department of Agriculture. The group also builds partnerships with timber companies, communities, NGOs, and agencies. While its conserved acreage is modest in Northwest Montana, its true impact is in restoring habitat, reducing risks for bears and people, and coordinating large landscape conservation. Developed areas present real challenges for bears. Doug Chadwick says, "Grizzlies run against land use barriers and get stuck where they are. That's why animals disappear on islands, because they've got no place to go. We're building corridors between the last remaining strongholds so they can survive. Vital Ground is making that real. We're helping to save Nature on a large scale and connected scale." He pauses. "That'll do."

Flathead Land Trust

Glacier National Park is the essential source of the Flathead River. Its glaciers, snowfields, and snowpacks sustain the North and Middle Forks, with the Bob Marshall and Great Bear Wilderness areas giving birth to the South Fork. All three are part of the National Wild and Scenic Rivers System, but that designation does not directly conserve adjacent land. The North Fork runs along Glacier Park's boundary and is highly vulnerable to development. In the 1980s, the Flathead National Forest and The Nature Conservancy began acquiring conservation easements on riverside properties from the Canadian border to Polebridge. But back then, the broad Flathead Valley was left unprotected.

Family time

Abundant water and fertile soils led to successful homesteading, and the Flathead Valley remains one of the state's most productive farming landscapes. Cattle and dairy operations are interwoven with fields of alfalfa, barley, corn, flax, oats, peas, lentils, squash, and wheat. Mild climates along Flathead Lake allow the growing of specialty crops like cherries, grapes, raspberries, strawberries, lawn sod, canola, and mint. The valley is also rich in wildlife habitats for resident and migratory birds. The Whitefish Range, the Swan Range, and the distant peaks of Glacier encircle everything. It is the empire of mountain goats.

Beauty like that attracts development, and Flathead County is one of the fastest growing in Montana. In 1970 the population was 39,000, and by 2020 it was more than 112,000, nearly tripling in 50 years. Tourism is a major economic sector, and some visitors move here to enjoy Nature year-round. One of Montana's ironies is that protected scenic vistas, agricultural land, and natural habitat lead to land subdivision and development. This once again leads to the Amenity Trap and conservationists have no choice but to do more.

Local residents sought the advice of The Nature Conservancy about how to proceed. In 1985, the Flathead Land Trust (FLT) was formed with a goal of conserving "working farms and ranches, critical habitats, and expanding recreation opportunities." In 1988, Alice Sowerwine donated the first easement to FLT, 157 acres on Fennon Slough. David Sowerwine said, "Hopefully, there will be a long series of future generations who will live in, and care for, and call the Flathead Valley 'home.'" State lands were added, and the Owen Sowerwine Natural Area now covers a square mile of bird heaven managed by Flathead Audubon and

FLATHEAD VALLEY
2021
LAND CONSERVATION STATUS 2021
Conserved Public Lands
Public Lands
Private Lands
Tribal Lands
Conserved Private Lands 1.5%
33.5%
33.1%
10.5%
21.4%
CANADA
MONTANA
GLACIER
NATIONAL
PARK
FLATHEAD
NATIONAL
FOREST
GREAT
BEAR
WILDERNESS
BOB
MARSHALL
WILDERNESS
FLATHEAD
RESERVATION
MISSION
MTNS
TRIBAL
WILDERNESS
STILLWATER
STATE
FOREST
COAL CK.
STATE FOREST
KOOTENAI
NATIONAL
FOREST
CORAM
EXP.
FOREST
RAY KUHNS
W.M.A.
WILD HORSE IS.
STATE PARK
SWAN
RIVER
ST. FOREST
NINEPIPE
N.W.R.
BISON
RANGE
Kintla Lake
Bowman Lake
Logging Lake
Lake McDonald
Duck Lake
Milk River
Saint Mary Lake
Dickey Lake
Stillwater Lakes
Whitefish Lake
Tally Lake
Ashley Lake
McGregor Lake
Lake Mary Ronan
Hungry Horse Reservoir
Swan Lake
South Fork
Flathead Lake
Pablo Reservoir
Ninepipe Reservoir
Flathead River
Big Salmon Lake
Flathead River
Gibson Reservoir
Holland Lake
Swan River
L. Inez
Seeley Lake
Clearwater River
Placid Lake
Jocko River
Clark Fork River
Whitefish R.
Flathead R.
Whitefish
Columbia Falls
Kalispell
Somers
Bigfork
Lakeside
Polson
Pablo
Hot Springs
Plains
Saint Ignatius
Superior
Arlee
Evaro
Seeley Lake
93
89
2
35
83
28
200
90
0 5 10 20 Miles

Montana Audubon. Projects grew thanks to the expertise of staff members like Laura Katzman, Ryan Hunter, Jen Guse, and executive director Paul Travis. A 284-acre farm was saved by the Cummings family, a fifth-generation legacy. Allan and Sallie Gratch protected a 160-acre piece with a historical cabin. Glenn and Hazel Johnston conserved a mile of the Flathead River, including a tall stand of ponderosa pines and endless "sacred rocks," as Glenn calls them. As is typical across Montana, opportunistic easements evolved into a strategic vision.

FLT is guided by its River to Lake Initiative, focused on receiving conservation easements on wildlife-rich agricultural properties beside the Flathead River. The river meanders across the valley, forming oxbow lakes and sloughs next to pothole lakes created when the Ice Age glacier retreated. Farms coexist with wildlife and enhance habitats in many ways, but housing development threatens that balance. Food production, scenic beauty, and culture are diminished with every subdivision and water quality suffers. Groundwater depth is shallow—about 20 feet—so septic tanks send bacteria into waterways, degrading aquatic habitats. Loss magnifies loss. Salish people call Flathead Lake "Broad Water," and it remains ecologically critical for a constellation of species, especially waterbirds. Up to 200 tundra swans overwinter here with ducks, cormorants, and a choir of divers and dabblers.

FLT holds conservation easements on more than 13,000 acres across the county. Most of that is within a mosaic of wetlands and farms. Another 4,000 acres were saved by FLT in partnership with other NGOs and agencies. As always, the Montana Land Reliance holds easements on thousands of acres—a gold-standard player everywhere in the state. The Nature Conservancy is also a mainstay in the North Fork and across the region. Saving the Big Sky is a shared commitment.

The Flathead Indian Reservation

This landscape is included on the preceding map. The Mission Mountains Tribal Wilderness, the Bison Range, Ninepipe National Wildlife Refuge, and a few conservation easements are shown. However, the next chapter includes a more detailed map and text created by the Confederated Salish and Kootenai Tribes to display their expansive sense of what "land conservation" means.

Whitefish Mountain

Spending a cold, sunny day carving tight turns on downhill skis or wide bends on telemark boards is how winter is celebrated in Montana. Bears hibernate and big game species move down to low-elevation winter ranges. The Whitefish Mountain Resort operates when wildlife are safely away or hibernating and the land is blanketed in deep snow. Hotels, houses, and condos cluster at the base.

A forester and photographer named Danny On loved this place back when it was called Big Mountain. Danny served in the 101st Airborne Division during World War II, jumping into Belgium during the Battle of the Bulge. He was severely wounded by shrapnel, but upon recovering, he earned his bachelor's and master's degrees in forestry at the University of Montana. During those college years, Danny spent summers fighting fires as the first Asian American smokejumper in American history. He was famous for his "long-delay" jumps—waiting until the last second to deploy his parachute. Danny spent the following decades working as a silviculturist for the Flathead National Forest, helping shift the agency into more sustainable harvesting practices. He adored Montana, exploring its ecosystems, taking photos, and publishing books encouraging us to conserve. On a snowy January day in 1979, Danny was skiing alone on Whitefish Mountain when he fell into a tree well, a depression ringed by deep snow. His body was found the next day, trapped headfirst in a drift. He was just 54.

Danny On's many friends wanted to create a monument honoring him, including Wes Pengelly, a fellow veteran and wildlife biology professor at the University of Montana. Instead of a statue, they chose Nature. Today, the 3.8-mile Danny On Memorial National Recreation Trail leads from the base to the summit—2,200 feet of vertical gain. During summer, you climb through forests and grassy ski runs full of blooming beargrass. The panorama from the top takes in the Cabinet Mountains, Whitefish and Flathead Lakes, Glacier Park, and a green sea of forestlands. From this height you almost don't notice all the development in the Flathead Valley. To the north, this part of Whitefish Mountain is protected by a timberland easement held by Montana Fish, Wildlife and Parks. The F. H. Stoltze Land and Lumber Company agreed to keep 3,000 acres in working shape forever. Danny On would be gratified.

Western Red Cedars

A remnant stand of 1,000-year old trees is protected south of Troy. It contains some of the last old-growth western red cedars in the region that somehow escaped logging. This shady, lush habitat is called the Lower Ross Creek Research Natural Area and it's managed by the US Forest Service. Genetic wisdom is protected and when restoration is needed elsewhere, we know where the seeds are. Ross Creek is mirrored in Glacier National Park along Avalanche Creek's Trail of the Cedars. Geographic distance provides evolutionary safety, but in the Middle East, the Cedars of Lebanon were nearly lost to logging. Solomon used cedar wood to build the temple in Jerusalem, and then Phoenician and Roman shipbuilders overharvested the tree. The "Cedar of God" is mentioned 103 times in the Bible. In 1998, a surviving stand in Lebanon was declared a UNESCO World Heritage Site. It's protected by armed guards.

In Northwest Montana, we protect Nature in many ways—conservation easements, parks, wilderness, land purchases, and wise timber management. We cherish waterfalls, ranches, farms, working forests, and wildlife habitats. Guards are not needed because we agree to honor what has been decided, in spite of cultural and political differences. This shows that ideological diversity can be good for biological diversity when we respect each other. There is strength in the mix and hope in the outcomes.

FURTHER READING

Along the Trail: A Photographic Essay of Glacier National Park and the Northern Rocky Mountains. 1980. David Sumner. Danny On, photographer. Lowell Press.

Flathead Land Trust. www.flatheadlandtrust.org.

Four-Fifths a Grizzly: A New Perspective on Nature That Just Might Save Us All. 2021. Douglas Chadwick. Patagonia.

Lost Trail National Wildlife Refuge. www.fws.gov.

Sustainable Forestry Initiative. www.forests.org.

CHAPTER 14

The Flathead Reservation

Introduction

by Bruce Bugbee, Robert Kiesling, and John Wright, with contributions from CSKT staff

In this chapter, we are honored to offer the Confederated Salish and Kootenai Tribes' (CSKT) own telling of their remarkable conservation story. It is a story of the tribes' relationship with the land—what they have cherished from time immemorial, and what they have protected and restored in recent decades by implementing traditional cultural values with the tools of modern governance and science.

What follows touches on only some of the many conservation efforts that have unfolded on the Flathead Reservation. The CSKT have conveyed their accomplishments more fully in a wide range of materials, some of which are referenced under "Further Reading" at the conclusion of this chapter.

Of the twelve tribes and seven Indian reservations in Montana, the CSKT and the Flathead Reservation face unique challenges. The reservation sits in the middle of the US Highway 93 corridor, stretching from the Glacier National Park area and the Flathead Valley in the north to Missoula and the Bitterroot Valley in the south. It is one of the fastest-growing regions in the state, besieged by ever-intensifying pressure from developers. For the CSKT, the problem is compounded by the US government having taken unilateral action more than a century ago, in violation of the 1855 Treaty of Hellgate, to make reservation lands available to non-Indian ownership and development. The CSKT have responded to the loss of land and jurisdictional challenges with resilience and innovation.

Before we offer their story, we want to offer a glimpse of the some of the good work being done by other Indigenous nations across the state:

- The Blackfeet have formed their own land trust. They are creating a large tribal bison herd, blessed with wild bulls and cows. Some of their story is included in chapter 9, "The Rocky Mountain Front."
- The Crow succeeded in having the Forest Service recommend that the Crazy Mountains be declared a wilderness area. The range—known in Crow as Awaxaawippíia—lies more than a hundred miles from the reservation borders, but the tribe's connection to the Crazies remains eternal. Rose Williamson (Crow) says, "When we look to that mountain, we see its spiritual strength and it would be nice if it was left untouched, not exploited, just left alone. Up there it cleanses not just your body, it cleanses your spirit. When you're done, you come out clean."

  The Crow have also worked to ensure preservation of the Bighorn Canyon National Recreation Area, which lies within the reservation and is another place of sacred meaning. Crow people tell the story of a boy lost in the canyon who was saved by seven bighorn sheep led by Big Metal, the chief of that herd. The tribe says it "casts a spell over those who would let it. For this is the Bighorn, and as the Bighorn it shall survive." Today, thousands of visitors come to this conserved place.
- The Northern Cheyenne, one of ten bands of the Cheyenne Nation spread across the Great Plains from southern Colorado to the sacred Black Hills of South Dakota, are the first Native nation in the United States to secure Class 1 Air Quality authority, the same level of protection as exists in national parks and monuments. In 1977, the Environmental Protection Agency granted the tribe that designation, providing the Northern Cheyenne with a powerful tool to protect tribal members from pollution caused

by coal development and power plants. (The EPA later designated the Flathead and Fort Peck Reservations as Class 1).

- The tribal governments of the Fort Belknap and Fort Peck Reservations, home to the Gros Ventre, Assiniboine, and Sioux, have allied with the American Prairie organization to return bison to an isolated Great Plains landscape, a story told in chapter 16, "The Prairie." And they have been crucial partners in the Bureau of Land Management's "Seeds of Success" prairie restoration program, which combines Western science and Indigenous knowledge. The tribes are helping collect "sacred seeds" of green needlegrass, prairie Junegrass, and blue grama. Savannah Buckman Spottingbird says of the program, "It gave me peace and happiness. Lots of adults on the rez think there is no coming back, that nothing is ever going to change. But change is within ourselves." Fort Belknap Indian Community elder Donovan Archambault adds, "For too many generations Indigenous people sat in silence and sorrow as they witnessed the gutting and rape of their beloved Mother Earth. The grassland restoration project is a positive step in the right direction. It is time to regenerate a system which our ancestors lived by for generations. Take care of your mother and she will take care of you." (From 2021 to 2025, the program was supported and overseen by Missoula native Tracy Stone-Manning in her role as the Biden Administration's BLM Director.)
- Since 2021, the Chippewa Cree of the Rocky Boy's Reservation (named for tribal leader Ahsiniiwin, meaning Stone Child) have also pursued the restoration of bison, building a herd on 1,200 acres they call "The Pasture" near Box Elder. Twenty-one bison were brought in from the CSKT and the American Prairie project, with new calves arriving every spring. Jason Belcourt (Rocky Boy sustainability coordinator) says, "Every donation continues to strengthen our herd and helps revitalize our culture."
- The Little Shell of the Chippewa are a federally recognized tribe of Ojibwe people, with a tribal council and offices in Great Falls, but they do not yet have a reservation in Montana. As Chris La Tray conveys in his fine book, *Becoming Little Shell: A Landless Indian's Journey Home*, the Little Shell continue their long quest for a sovereign reservation and the opportunity to manage part of their homeland according to their cultural values and social needs.

In the pages that follow, the Confederated Salish and Kootenai Tribes relate some aspects of their own conservation story.

Continuance and Revitalization: Our Conservation Story

by staff members of the Confederated Salish & Kootenai Tribes, with contributions from Bruce Bugbee, Robert Kiesling, and Jack Wright

In 2019, the Director of the Séliš-Ql̓ispé Culture Committee, Antoine Incashola Sr., spoke at a meeting of the Climate Change Advisory Council of the Confederated Salish and Kootenai Tribes. He reminded us that "at the beginning of time, the animals that were here before us left us a near-perfect world. . . . And the elders always say that if you don't take care of what you've got, those gifts are going to leave."

This perspective has always stood at the heart of our way of life—and for the past half-century, it has formed the foundation of the CSKT conservation work we describe in this chapter.

For thousands of years, the mountains and valleys of what is now western and central Montana and adjacent areas have been home to our people: the Séliš (Salish or "Flathead"), Ql̓ispé (Upper Kalispel or "Pend d'Oreille"), and Ksanka (Kootenai) Nations. For the vast majority of that time, we lived well as hunters, fishers, and gatherers, moving with the seasons, drawing from a profound ecological understanding of our vast homelands. We lived in close tribal communities, conducting most activities together for the well-being of the people as a whole.

This is the way of life shown in the beginning by Coyote: a way of respect for one another, and for the lands and waters we have been given. Our

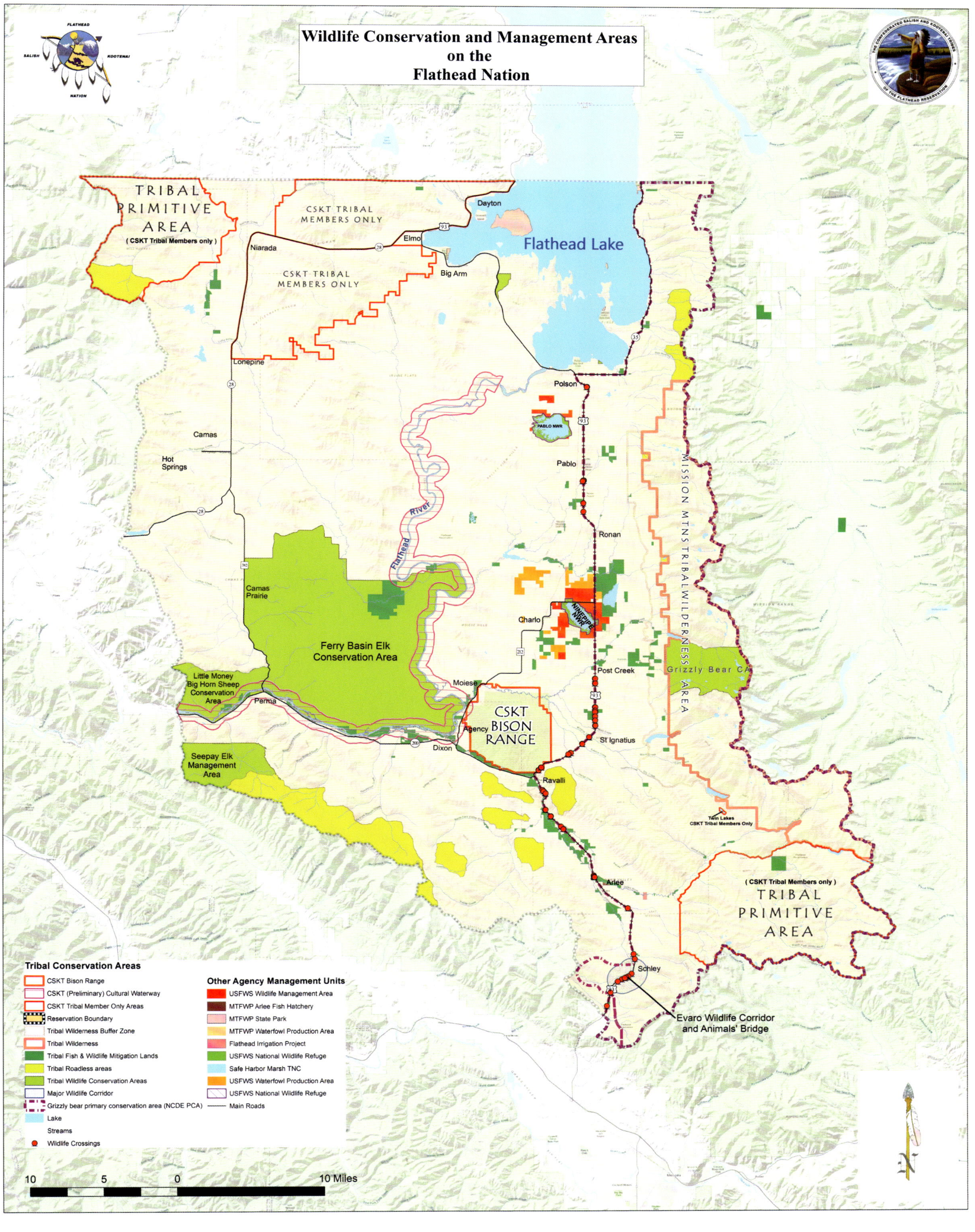
Wildlife Conservation and Management Areas
on the
Flathead Nation
TRIBAL
PRIMITIVE
AREA
(CSKT Tribal Members only)
CSKT TRIBAL
MEMBERS ONLY
Flathead Lake
Dayton
Elmo
Niarada
Big Arm
Lonepine
Polson
Camas
Hot Springs
Pablo
Ronan
Flathead River
Camas Prairie
Charlo
Ferry Basin Elk
Conservation Area
Little Money
Big Horn Sheep
Conservation
Area
Perma
Moiese
Post Creek
CSKT
BISON
RANGE
Agency
Dixon
St Ignatius
Seepay Elk
Management
Area
Ravalli
Arlee
MISSION MTNS TRIBAL WILDERNESS AREA
Grizzly Bear CA
Twin Lakes
CSKT Tribal Members Only
(CSKT Tribal Members only)
TRIBAL
PRIMITIVE
AREA
Schley
Evaro Wildlife Corridor
and Animals' Bridge
Tribal Conservation Areas
CSKT Bison Range
CSKT (Preliminary) Cultural Waterway
CSKT Tribal Member Only Areas
Reservation Boundary
Tribal Wilderness Buffer Zone
Tribal Wilderness
Tribal Fish & Wildlife Mitigation Lands
Tribal Roadless areas
Tribal Wildlife Conservation Areas
Major Wildlife Corridor
Grizzly bear primary conservation area (NCDE PCA)
Lake
Streams
Wildlife Crossings
Other Agency Management Units
USFWS Wildlife Management Area
MTFWP Arlee Fish Hatchery
MTFWP State Park
MTFWP Waterfowl Production Area
Flathead Irrigation Project
USFWS National Wildlife Refuge
Safe Harbor Marsh TNC
USFWS Waterfowl Production Area
USFWS National Wildlife Refuge
Main Roads
10
5
0
10 Miles

Creation stories teach us that our lives are only possible because of the sacrifice and generosity of the plants and animals. So before harvesting anything, we offer a gift. We give thanks and pray that they will continue to provide for us. Upper Kalispel elder Pat Pierre reminded us many times that we must give before we take, and we must never waste.

During recent centuries, we have endured the invasion and transformation of our homelands. But our people resisted, and over the past half-century, within our remaining sovereign territory—within the Flathead Reservation—we have protected, restored, and revitalized our sovereignty, cultural ways, and natural resources.

In telling this story, it is important to convey that for us, **natural resources *are* cultural resources.** There is no difference between them. The songs, dances, stories, and our understandings of the origins of human life—all are gifts from the animals, plants, lands, and waters. This basic understanding undergirds CSKT cultures. And it also undergirds the entire CSKT conservation record.

This chapter presents just a brief introduction to our cultural-environmental work. We touch on seven examples. Other CSKT initiatives and programs could merit whole chapters in themselves, but are not discussed here; for example, the Tribal Forestry Program, guided in large part by its revolutionary 2000 Forest Management Plan, which shifted our long-term goals from commodity production to the restoration of former forest landscapes, including reestablishing pre-1850 fire regimes through active prescribed fire programs. Neither does this chapter tell the stories of the CSKT Air Quality and Water Quality Programs; the major stream restoration work under the CSKT Water Compact; the Underground Storage Tank Program; or the Pesticides Program, among others.

Our people have been here a long time. Tribal creation stories contain memories of the last Ice Age, echoed in archaeological sites dated to almost 13,000 years ago, about the time of the last draining of Glacial Lake Missoula. If we take just the most recent 9,000 years of our history and compress it into a single 24-hour day, 1800 would arrive around 11:29 p.m.—about half an hour before midnight. In the context of tribal tenure, all of the change that has come to our homelands and our people since then is very recent and very sudden.

In 1855, US officials and tribal leaders "negotiated" the Treaty of Hellgate, which in many ways defined and shaped the subsequent political, legal—and environmental—landscape of western Montana. Poor translation limited how much the chiefs could understand, but according to the terms of the treaty, we ceded most of our territories to the United States. We reserved tribal rights to hunt, fish, and gather plants on those ceded lands, and also reserved from cession a small fraction of our vast homelands, including the Flathead Reservation, as sovereign homelands. The treaty guaranteed the reservation for "the exclusive use and benefit" of CSKT people.

Less than fifty years later, Congress broke that promise with passage of the Flathead Allotment Act, despite nearly unanimous tribal opposition and unrelenting efforts by tribal leaders to stop the bill. The 1904 law made large swaths of the Flathead Reservation available to non-Indian homesteaders, who quickly took control of much of the agriculturally and commercially valuable land. For the CSKT, allotment was the single most devastating policy ever implemented by the federal government, bringing the dispossession of over half a million acres, the radical weakening of our sovereignty, and the transformation of the reservation's economy and culture. The result was the complex, sometimes conflicting mix of landownership and jurisdiction that characterizes the Flathead Reservation to this day.

In 1934, FDR's Indian Reorganization Act finally ended the disastrous allotment policy and gave us the opportunity to reconstitute as the Confederated Salish and Kootenai Tribes. We began rebuilding our governing capacity. We also started reacquiring land; today, the CSKT own over two-thirds of our 1.3-million-acre reservation.

In the mid-1970s, our Tribal Council took a crucial step by establishing Séliš-Ql̓ispé and Ksanka Culture Committees. Arising from the concern of tribal elders over the loss of cultural knowledge in the community, the culture committees are charged with preserving, protecting, and perpetuating the cultures, languages, and history of the Salish, Kalispel, and Kootenai people. With the guidance and direction of Elders Cultural

Advisory Councils, the culture committees helped restore traditional cultural values—including respect for the environment—to the heart of CSKT operations.

The following decades saw the development of many CSKT departments and programs, including the Natural Resources Department. Terry Tanner, who worked for both the Séliš-Ql̓ispé Culture Committee and the Natural Resources Department, said, "We honor our ancestors through our stewardship of the land and by maintaining and exercising the rights we kept in our treaty. This is our generational responsibility that we grow up with as Indian people."

Seven examples illustrate that generational responsibility in action.

THE MISSION MOUNTAINS TRIBAL WILDERNESS

In 1982, the CSKT Tribal Council passed an ordinance establishing the Mission Mountains Tribal Wilderness, encompassing some 92,000 acres—the first tribally designated wilderness in the United States.

Like federal wilderness areas, the CSKT wilderness excludes roads, motorized vehicles, or other machines. But in other ways, the Missions are managed as a uniquely tribal conservation area, springing from the mountains' deep cultural, spiritual, and historical importance for our people. While federal wildernesses are defined by the minimization and sometimes the erasure of human presence, the Missions are held by the CSKT as a haven for tribal cultural continuance. This was expressed eloquently by former Séliš-Ql̓ispé Culture Committee Director Clarence Woodcock: "The Mission Mountains have served as a guide, passageway, fortification, and vision-seeking grounds, as well as a place to gather medicinal herbs, roots, and a place to hunt for food for the Pend d'Oreille, Kootenai, and Salish people. They have become for us . . . sacred ground. Ground that should not be disturbed or marred."

For much of the twentieth century, the CSKT had to fend off attempts by the Bureau of Indian

Mission Mountains on the Reservation

Affairs to log the mountains. In the 1930s, shortly after our adoption of a new Tribal Constitution, the first Tribal Council tried to permanently protect the Missions, but the BIA ignored the request. In 1974, the BIA proposed a timber sale that would have clear-cut some 2,000 acres in the heart of the range near Ashley Lakes. When the Tribal Council met to consider the potentially lucrative plan, Councilman Thomas "Bearhead" Swaney invited three highly respected tribal elders to speak: Annie Pierre, Louise McDonald, and Christine Woodcock, known to the community as the Three Yayas (grandmothers). They told the Tribal Council that "the Mission Mountains were a treasure and that it was important not to destroy them in the short time we are here." Tribal educator Germaine White, who accompanied the Three Yayas, recalled that the council listened, thanked the Yayas, and waited for them to leave. They stayed. When the tribal chair asked whether they had anything more to say, one of the grandmothers said, "We'll just wait here until you vote." By a 6–2 margin, the Council took the unprecedented action of tabling the timber sale.

Over the next eight years, Tribal Council took a series of steps, including hiring the Wilderness Institute at the University of Montana to create boundaries and management guidelines. Council actions were supported and encouraged by tribal members such as Thurman Trosper, a retired US Forest Service supervisor and past president of The Wilderness Society, and Doug Allard, who was helping lead a newly formed group called the Save the Mission Mountains Committee.

The Mission Mountains Tribal Wilderness has a number of unique attributes, including a special grizzly bear management zone (closed to all use during certain periods of the year to protect bears feeding in high elevation areas) and what is called the "Buffer Zone" along the western base of the mountains—an area outside the official wilderness boundaries, but managed to protect the ecological and cultural integrity of the wilderness itself. An adjoining area to the south is designated as a tribal-member-only "Primitive Area"; although it is accessible by vehicle, it is otherwise protected and pristine. Areas abutting the wilderness on the north are designated as CSKT Roadless Areas. The ecological value of these adjoining areas, particularly for the connectivity of wildlife habitat, is further augmented by nearby off-reservation conservation lands, including the federal wilderness encompassing 73,877 acres on the east side of the Missions, and across the Swan Valley, the sprawling Bob Marshall Wilderness, covering over a million acres.

In 2022, the three yayas—Annie Pierre, Christine Woodcock, and Louise McDonald—were honored for their leadership with induction into the Montana Outdoor Hall of Fame. But their greatest legacy—the legacy of so many elders, cultural leaders, CSKT staff, and consultants—is the permanent protection of these spectacular, powerful mountains.

THE BISON RANGE

The relationship with q̓ʷeyq̓ʷay (buffalo) stands at the heart of our cultures and histories. It is a relationship of reverence and respect. The elders have told how our lives, human beings and buffalo, are intertwined. Our people have always sought to take no more than we need; to use all parts of the animal; to waste nothing. Throughout our history, we have worked to ensure that buffalo will be plentiful for all the generations to come.

In the years after the 1855 Hellgate Treaty, we could see that buffalo were declining, due primarily to non-Indian hunting. Intertribal conflicts over the dwindling resource were intensifying. In response, a Ql'ispé man named Atatic̓e? (Peregrine Falcon Robe) brought a profound idea before the chiefs: bringing buffalo back to establish a herd within the Flathead Reservation. His son, Susép Łatatí (Joseph Little Falcon Robe), later carried out his father's vision, capturing some orphaned calves and herding them west. In 1884, tribal members Michel Pablo and Charles Allard purchased the growing herd, eventually augmenting them with buffalo purchased from others. It became the largest remaining herd in the United States, thriving in the tribal lands along the lower Flathead River. Allard died in 1896; some of his buffalo were then sold to the Conrad family in Kalispell.

Over the next few years, CSKT bison conservation efforts were upended by the federal government. The Flathead Allotment Act of 1904 would

transfer many reservation lands to non-Indian homesteaders, eliminating much of the reservation's open range. US officials told Michel Pablo he had to get rid of his buffalo. No American buyers emerged, so in 1906, Pablo sold them to the Canadian government. He then organized buffalo roundups, which continued each year until 1912.

During this very same time period, wealthy New Yorkers formed the American Bison Society (ABS), whose honorary chair was Theodore Roosevelt (who as President signed the Flathead Allotment Act). In 1909, the ABS convinced Congress to expropriate reservation land from the Flathead Nation to form a National Bison Range. Elders have told how tribal leaders voiced their vehement opposition to this taking of prime hunting grounds, but US officials told us we had no choice in the matter.

The government, having forced the elimination of a tribal herd, now seized tribal land—some 18,882 acres—to form a herd under their control. The United States furthermore determined, unilaterally, the amount it would pay for the land—and then expended those funds to cover the administrative costs of making reservation lands available to non-Indian homesteaders. One act of dispossession funded another.

In further irony, the ABS then acquired 34 buffalo from the Conrad herd in Kalispell—descendants of the buffalo saved decades before by Little Falcon Robe—and donated them to the federal government to seed the National Bison Range. (Portions of Allard's buffalo had been sold to Howard Eaton, who later sold his animals to Yellowstone Park. Thus the buffalo originally saved by the tribes also helped seed the famed Yellowstone herd.)

Beginning in the 1990s, CSKT Tribal Council and staff pursued a variety of strategies to regain control of the Bison Range. This became a reality in 2020, when Congress passed Public Law 116-260, under which the Bison Range remains federal land but is now held in trust for the CSKT. After more than a century of dispossession, the CSKT is reconnecting with this land at the heart of the Flathead Reservation. Now managed by our award-winning Natural Resources Department, the Range is home to about 350 adult buffalo, with about 50 to 60 calves welcomed to the world each year. Visitors still have the opportunity to see bison rolling in dusty wallows, grazing in large herds, calves frolicking next to cows, bulls moving with quiet power across the land. The Range's loop road traverses places that now bear names rooted

Bull bison on the range

At the dedication of the CSKT Bison Range, May 21, 2022, Tribal Council Chair Tom McDonald stands with Deb Haaland (Laguna Pueblo), who served the Biden Administration from 2021 to 2025 as the first Indigenous Secretary of the Interior in American history. Photo by Tailyr Irvine (CSKT), courtesy CSKT.

in CSKT culture and history, such as Peregrine Falcon Robe's Home and Place Where You Watch Out for Something That's Coming—a high point offering expansive views of the prairie hills sloping down to St. Ignatius, and beyond, the snowy crest line of the Mission Mountains. Here, we will carry on for the generations yet to come our ancient relationship with q̓ʷeyq̓ʷay.

The CSKT herd is now helping other Indigenous nations replenish and repopulate their buffalo. And here at home, we are planning to establish a herd outside the Bison Range that will serve our Food Sovereignty Movement as a sustainable food source.

The full circle of the CSKT, buffalo, and the National Bison Range is encapsulated in the story of Big Medicine. On May 3, 1933, range rider John McDonald came across something rare and revered in our culture: a newborn white bison bull. Within days, tribal people conducted a ceremony to welcome this special gift from the Creator. Some came to call the buffalo "Big Medicine." After he died in 1959, CSKT chair Walter McDonald—John's brother—expressed the community's wishes that the white buffalo be preserved and kept on the Flathead Reservation. However, National Bison Range officials instead conveyed the hide to the Montana Historical Society. For half a century, the taxidermized buffalo stood on public display at the MHS museum in Helena. CSKT members, including many elders, made numerous journeys to visit Big Medicine and pray for his return. On October 20, 2022, those prayers were finally answered. At the urging of tribal leaders, the MHS Board of Directors voted unanimously to return the white buffalo to the CSKT Bison Range. For our people, this makes complete the restoration of the Bison Range. Vice-Chair Tom McDonald—Walter and John's grandnephew—said, "Big Medicine represents the past that has carried us forward to the present, and the work yet to be done to protect our identity, culture, and well-being into the future."

Sandhill liftoff

WILDLIFE MANAGEMENT AND CONSERVATION AREAS

The CSKT's Wildlife Program has garnered widespread recognition and awards for its achievements in carrying out the "protection, enhancement and management of terrestrial wildlife species and habitats" on the Flathead Reservation. The program's work can be seen in almost all the other sections of this chapter, including in the discussions of the Bison Range, the Mission Mountains Tribal Wilderness, Wetlands, and the Highway 93 Wildlife Crossings. CSKT wildlife biologists have been active in the recovery of federally listed threatened and endangered species of native wildlife, including bald eagles, peregrine falcons, northern gray wolves, grizzly bears, Canada lynx, and northern leopard frogs. Trumpeter swans—once locally extirpated—now thrive in the valley thanks to the CKST's reintroduction program.

The program has mapped five prime habitats for elk, bighorn, and grizzly bears within the reservation, including the Grizzly Bear Conservation Area (CA) in the Mission Mountains, the Ferry Basin and Seepay Elk CAs, and the Little Money and Hog Heaven Bighorn Sheep CAs.

CSKT staff have worked hard and with great expertise in managing and sustaining the Mission Mountains subpopulation of grizzlies, who frequently venture out into the increasingly subdivided valley.

Again, this area of CSKT conservation work springs from a deep cultural foundation. Salish elder Johnny Arlee said, "These animals were here before us and we respect them in this way."

WETLANDS

The Flathead Reservation is home to an astounding number and variety of wetlands, including Flathead Lake, the lower Flathead and Little Bitterroot Rivers, major streams issuing out of the Mission Mountains, and hundreds of glacial pothole ponds in and adjacent to the Pablo and Ninepipe National Wildlife Refuges. As Vice-Chair Tom McDonald has said, these are "places of special abundance." CSKT efforts to protect wetlands again spring from the cultural importance of the resources found there: plant foods, medicines, and materials, including edible valerian, gooseberry, cattail, and bulrush, among many others; and animals including turtles, frogs, muskrats, bald and golden eagles, great blue herons, snipe, bitterns, and waterfowl.

The CSKT maps and monitors the health of reservation wetlands, working to prevent or minimize draining, filling, and pollution. Other CSKT programs track and control invasive species. The CSKT and the US Fish and Wildlife Service work together in numerous ways, including installing cattle-exclusion fencing around wetlands. And under the historic Water Compact, CSKT staff are addressing impacts from the Flathead Indian Irrigation Project, including restoring numerous vital wetlands.

CSKT FISHERIES AND JOCKO RIVER RESTORATION

Across the Flathead Reservation, the CSKT Fisheries Program has implemented a range of programs to fulfill its stated mission of "restoring, fostering, and maintaining wild, self-sustaining fish populations to meet cultural, subsistence, and recreational needs." From surveying fish populations to restoring habitat, from meeting the threat of aquatic invasive species to addressing the unique biological and jurisdictional challenges of Flathead Lake, the program has a long record of accomplishment.

These initiatives serve a resource whose importance in tribal life and history has been powerfully conveyed by elders. The CSKT's award-winning environmental education project, *Explore the River: Bull Trout, Tribal People, and the Jocko River* (2011), reveals the critical role played by fishing and the abundant fisheries of CSKT territories—how the steady availability of fresh, high-quality aquatic protein functioned as a key element when some other traditional food sources were more seasonal or sporadic. Salish educator Germaine White said, "We knew where the streams were, where there were upwellings, where there were a lot of fish. We knew how to deal with seasonal abundance and we knew how to take advantage of this high quality protein source." Upper

Kalispel elder Mitch Smallsalmon conveyed how the fisheries were a critical underpinning of tribal well-being: *Kʷem̓t šey̓ še nk̓ʷúlexʷ qe sqʷyúlexʷ łiʔe l sewłkʷ.* "By that, we were wealthy—from the water."

Explore the River explains that no fish holds more importance, culturally or materially, than the greatest of all the native species: *aáycčst*, or bull trout. They swarmed in unimaginable numbers through tribal waters, with the biggest adfluvial fish reaching over three feet in length.

But as with many other traditional foods, bull trout have been devastated by the transformation of tribal territories over the past century and a half. Even within the Flathead Reservation, the species became imperiled in recent decades. Bull trout disappeared from Mission Creek some twenty years ago, and for the past fifteen years, only about 3,000 adults have remained in Flathead Lake, their numbers decimated by non-native lake trout.

The upper Jocko River remained one of the few reservation strongholds for bull trout. The lower Jocko was a different story, where major degradation began with construction of the Northern Pacific Railroad in 1883. CSKT biologists found that since 1910, the lower river "had been substantially disturbed by agriculture, irrigation withdrawals, livestock grazing, transportation infrastructure, and residential/commercial development."

Now the river is becoming an excellent example of ecosystem recovery. In 2008, the Jocko Restoration Project—funded by a Superfund settlement, with additional support from the Bureau of Reclamation, the CSKT's Séliš Ksanka Ql'ispé Dam, and the CSKT Water Compact—set out to heal the damaged waterway's lower 22 miles. The Jocko River Master Plan mixed "active restoration" (physically repairing the most degraded reaches of the river) with "passive restoration" (modifying land-use practices). CSKT staff have reconstructed and rehabilitated canals, restored cottonwood ecosystems, removed cattle from sensitive areas, removed riprap, rebuilt meanders and pools, improved fish ladders, and returned floodplains to their natural state. We have also reacquired the majority of lands along the river. At the same time, SKQ Dam operations were changed to reduce impacts to the lower Flathead River. As of 2024, the whole Jocko system is seeing a rebound in native fish populations, including westslope trout and bull trout. The success in the Jocko bodes well for other CSKT efforts in bull trout recovery, including plans to reintroduce the species to Mission Creek, and the ongoing effort to reduce lake trout in Flathead Lake.

THE ANIMALS' BRIDGE

North of Evaro, a high curving bridge spans US Highway 93. It's not meant for cars or people. A Salish place-name sign identifies it as Xʷixʷey̓úł Nx̣lew̓s—the Animals' Bridge. The structure provides safe wildlife passage over the busy highway, thereby protecting both motorists and animals from collisions. Fences guide animals to the bridge. North of there, culvert underpasses and bridging structures, also with funneling fences, provide routes for animal passage underneath US 93. Between the reservation boundary and Polson, a total of 42 crossing structures of various designs are now in place. More will soon be built in the final section of the road awaiting reconstruction, the wetland-rich area of the Mission Valley known as Ninepipes. CSKT's Natural Resources Department has shared photos of bears, mountain lions, elk, deer, and coyotes safely moving through the structures. Tribal Vice-Chair Tom McDonald reminded us that "wildlife need movement corridors. Animals need a chance to adapt, especially when you think about climate change that's hitting us so hard. You really have to step back and give wildlife some space."

The wildlife structures were built as part of US 93's reconstruction during the early 2000s. They constitute a monumental achievement in rural highway design in the United States, protecting and restoring habitat connectivity by providing safe wildlife movement across the heavily trafficked highway corridor.

The history of the wildlife crossings is part of the larger, decades-long struggle over US 93, which involved environmental issues rarely addressed with any success: the threat of uncontrolled growth and sprawl, the role of highway

Animals' bridge

expansion in accelerating those problems, and ultimately, the dominant society's disastrous assumption that we can accommodate and feed infinite growth on a finite planet. When the Montana Department of Transportation (MDT) first proposed a dramatic expansion of the highway in the late 1980s, CSKT and many other local residents saw it as a direct threat to the character of the Flathead Reservation's cultural and natural environment. That radical widening of the pavement would have further intensified the already daunting problem of uncontrolled growth spreading northward from the sprawling city of Missoula. Suburbanization and traffic volume and speeds are particular concerns for the CSKT as we work to regain our land base within the reservation, and with it, the continuance of our social and cultural community. Ultimately, the Tribal Council saw the MDT plan as posing a clear and present danger to tribal sovereignty itself.

For the CSKT and especially tribal elders, the deep tribal ethic of living within limits, of not taking too much, underlay our consistent opposition to the MDT plan and our drive to find a less destructive solution to address safety concerns. With critical support from the Flathead Resource Organization ("FRO," a long-standing local environmental group comprised of both tribal members and non-Indians), the CSKT thus took a courageous stand on the biggest issue, lane configuration, opposing the MDT plans for a four- and five-lane undivided highway and instead supporting a "Super Two" lane design (two lanes with alternating passing lanes, consolidated entrances, and widened shoulders). In the historic 1996 Record of Decision on the project's Environmental Impact Statement, the Clinton Administration's Federal Highway Administration (FHWA) backed CSKT concerns. For the first time, FHWA affirmed that highway expansion, in certain contexts, doesn't just accommodate increasing traffic, but also fuels and accelerates uncontrolled growth and sprawl. Recognizing the federal government's "trust responsibility" to the CSKT as a tribal nation, FHWA informed MDT and CSKT that it would only fund the project once the two sides reached agreement on a design. The FHWA decision finally forced MDT to negotiate.

The issue began gaining national attention. In 1997, CSKT and FRO successfully nominated the entire Flathead Reservation to the National Trust for Historic Preservation's annual list of America's 11 Most Endangered Places. CSKT and MDT finally reached agreement in 2000, ending up with a highway that garnered major awards for "context sensitive design." The new design was not only far less environmentally destructive, but also highly effective in reducing accidents and fatalities on the heavily traveled road, as confirmed in subsequent safety audits.

The "context sensitivity" of the highway's design included wildlife crossing structures as well as Salish and Kootenai language place-name signs. It is important to understand that for the MDT, these were "mitigation" measures they only agreed to in order to gain CSKT compromise on the bigger issue of lane configuration. Neither the wildlife crossings nor the place-name signs would exist today if CSKT elders, Tribal Council, and staff had not taken their courageous stand against MDT's multi-lane strip highway. That stance gave the CSKT leverage.

Both the wildlife crossings and the place-name signs were aimed not just at limiting the destruction caused by the highway's reconstruction, but also healing some of the existing damage. They are another example of how CSKT has pursued not only conservation but also restoration. The crossings have enhanced and restored the connectivity of wildlife habitat, and the signs have enhanced and restored travelers' awareness of the peoples and cultures that have shaped this place from the beginning of human time.

THE CULTURAL WATERWAYS ORDINANCE

In 1938, the Montana Power Company completed construction of a dam on the lower Flathead River, about eight miles below the outlet of Flathead Lake. Named after MPC president Frank Kerr, the dam and its operation caused significant ecological harm to the lake and river, cultural landscapes of the highest importance for the CSKT. In the 1991 film *The Place of the Falling Waters*, elders convey that many tribal people opposed the dam but had no voice in the matter. In the decades that followed, the remaining free-flowing sections of the river were repeatedly threatened by proposals for additional dams. Opposition from our leadership, elders, and the wider community managed to stop those projects. Had they been built, they would have flooded almost the entire river down to its confluence with the Clark Fork near Paradise, Montana. CSKT leaders and staff also had to fight off many other threats, including BIA proposals to log the old growth riparian pines, proposed subdivision developments, and gravel and rock mines.

For nearly 75 years, the dam and its revenue were controlled by Montana Power. But in 2012, CSKT Tribal Council decided to act on the opportunity provided under a 1984 licensing agreement: the CSKT purchased the dam, thus becoming the first Native nation in the United States to be the sole owner of a major hydroelectric facility.

The CSKT also decided that even as we became owners of the dam, we would ensure that the rest of the lower Flathead would be protected. This followed decades of work by CSKT cultural leaders and resource specialists to protect and restore the river's fish and wildlife, reacquire riparian lands, develop management plans, and educate the membership and the public, including hosting the annual River Honoring, one of Montana's largest outdoor environmental education events. All of those efforts culminated on August 10, 2021, when the Tribal Council passed the CSKT Cultural Waterways Ordinance. The ordinance states, "Water is sacred. . . . Our people carry a spiritual obligation to protect our clean, abundant waters, and to maintain the free-flowing nature of those stretches of the river that still remain undammed for the generations to come."

The Tribal Council further designated the Lower Flathead River as the first CSKT Cultural Waterway, "to protect and preserve the Lower Flathead River Cultural Waterway in perpetuity as a Tribal traditional cultural sanctuary and an area of land preserved in a generally natural condition." The ordinance states, "There shall be no additional dams constructed on the Lower Flathead River."

Just as the CSKT became the first Native nation to establish the tribal equivalent of a federal wilderness area when they set aside the Mission Mountains Tribal Wilderness, so the tribes established the tribal equivalent of a Wild and Scenic River. But like the tribal wilderness, CSKT cultural waterway designations are framed first and foremost around the protection and continuance of tribal cultures and cultural values—a powerful reflection of how, for the CSKT, relationships of respect with the waters, the lands, the plants, and the animals stand at the heart of who the Salish, Kalispel, and Kootenai people are.

For the Generations to Come

The seven CSKT projects discussed in this chapter demonstrate how we are finding new ways to bring forward our traditional principles of respect for and reciprocity with the environment. Those deep cultural values govern our resource management decisions.

We know that new and even more daunting challenges lie ahead. For millennia, our people have adjusted successfully to climactic shifts. But now all of us are faced with change of a different magnitude. The climate crisis threatens to unravel all that sustains us and all that we are working so hard to pass on to the generations to come. This is why, in 2016, the CSKT became the first US tribal nation to adopt a Climate Change Strategic Plan. We are doing our best to safeguard our people and our homelands and to continue bringing forward to the world the wisdom and the example of our ancestors. Their example—our history—helps us realize that catastrophe is not inevitable, that we human beings are capable of living in a way that nurtures rather than destroys our only home.

We return to what Tony Incashola Sr. told us at a CSKT climate gathering in 2019: "At the beginning of time, the animals that were here before us left us a near-perfect world. . . . And the elders always say that if you don't take care of what you've got, those gifts are going to leave. . . . Time is getting shorter. . . . The thing that I remember . . . is our elders always used to say, 'Don't give up. Never give up.'"

We are trying to follow that guidance: to never give up, to continue our efforts to protect and restore what we have been given. As elder Pat Pierre told us, "We have generations yet coming. We need to work on it every day."

FURTHER READING

All websites accessed 2025-01-14.

Apsáalooke Nation (Crow Tribe). www.crow-nsn.gov.

Becoming Little Shell: A Landless Indian's Journey Home. 2024. Chris La Tray. Milkweed Editions.

Blackfeet Nation. www.blackfeetnation.com.

"Buffalo, the Séliš and Ql'ispé People and the Restoration of the Bison Range." 2023. Séliš-Ql̓ispé Culture Committee. www.pbs.org/kenburns /the -american-buffalo/ restoration-of-the-bison-range.

Confederated Salish and Kootenai Tribes (CSKT). www .cskt.org.

- Climate: www.csktclimate.org.
- CSKT Bison Range: www.bisonrange.org.
- Fisheries Program: https://cskt.org/natural -resources/fisheries-program/.
  - Includes: Jocko River Master Plan: Executive Summary. 2008.
- Forestry Department: https://cskt.org/forestry/.
- Forestry—Division of Fire: https://cskt.org /division-of-fire/.
- Natural Resources Department (NRD): https:// cskt.org/natural-resources/
- NRD Division of Fish, Wildlife, Recreation & Conservation: https://cskt.org/natural -resources/dfwrc/.
  - Includes: "Flathead Indian Reservation Fishing, Bird Hunting, and Recreation Regulations of the Confederated Salish and Kootenai Tribes and the Montana [Department of] Fish, Wildlife & Parks . . . for Persons Who Are Not Enrolled Members of the Tribes." 2024.
- NRD online educational resources: http:// fwrconline.csktnrd.org.
  - Includes: *Fire on the Land: Native People and Fire in the Northern Rockies*. 2007.
  - *Explore the River: Bull Trout, Tribal People, and the Jocko River*. 2011.
- Séliš-Ql'ispé Culture Committee (SQCC): www .csktsalish.org (in 2025, the primary site will be https://cskt.org/selis-ql̓ispe-culture -committee/).
  - Main page includes: "2022 Brochure and Guide to Available Educational Resources."
  - "Audio" tab includes: "Salish-Pend d'Oreille Placename Signs on Highway 93." 2011.
  - "Documents" tab includes:
    - *Aáy u Sqélixʷ: A History of Bull Trout and the Salish and Pend d'Oreille People. 2011. Booklet. Produced as part of the CSKT digital education project, Explore the River: Bull Trout, Tribal People, and the Jocko River.*
    - *Indigenous Peoples and Forests. 2020. Booklet. Produced for the Montana Forest Action Plan, drawing from essays produced for the CSKT digital education project, Fire on the Land: Native People and Fire in the Northern Rockies.*
    - "Q̓ʷeyq̓ʷay: *Buffalo and the Séliš & Ql'ispé People." 2022. Pamphlet.*
    - *Saving the World that Coyote Made: The Climate Crisis and Native People of the Northern Rockies. 2024. Booklet. Written by Thompson Smith for the "Living Landscapes Project," co-directed by Germaine White, David Rockwell, and Adrian Leighton.*
    - Sxʷúytis Smx̣e Nx̣lew̓s / *Grizzly Bear Tracks Bridge: Historical Background. 2022. Booklet.*
  - "Ethnogeography" tab includes:
    - Skʷskʷstúlexʷ / *Names Upon the Land: Ethnogeography of the Salish and Kalispel People: Introduction to A Portfolio and Maps and Signs. 2019. Booklet.*
    - "Sl̓l̓túlixʷs Séliš u Ql̓ispé: *Territories of the Salish, Kalispel, and Related Indigenous Nations." 2023. PDF.*

Coyote Stories of the Montana Salish. 1999. Salish Culture Committee and Johnny Arlee. Montana Historical Society Press.

Fort Belknap Indian Community, Home of the Nakoda and Aaniiih Nations. www.ftbelknap.org.

Fort Peck Assiniboine and Sioux Tribes. www .fortpecktribes.org.

"Habitat Restoration and Fisheries Programs Help Fish Thrive in Jocko River." 2019. Alyssa Kelly. *Char-Koosta News*, August 29, April 2019. https://www .charkoosta.com/news/cskt-habitat-restoration-and -fisheries-programs-help-fish-thrive-in-jocko-river /article_cc7274c0-ca80-11e9-baa5-d3e171c2b2e4 .html.

"Highway Expansion and Sprawl Threaten One of Montana's 'Last Best Places.'" 1997. Thompson Smith. *Forum Journal* (National Trust for Historic Preservation) 11, no. 4: 34–43.

In the Name of the Salish and Kootenai Nation: The 1855 Hell Gate Treaty and the Origin of the Flathead Reservation. 1996. Edited by Robert Bigart and Clarence Woodcock. Salish Kootenai College, Pablo, Montana.

In the Spirit of Atatiče?. 2018. Film. Directed by Daniel Glick. Confederated Salish & Kootenai Tribes. https://www.youtube.com/watch?v=S1WvkSN8zDQ.

"Interior Secretary Deb Haaland's Charged Mission of Healing." July 17, 2023. *Washington Post*. https://www.washingtonpost.com/lifestyle/2023/07/17/deb-haaland-road-to-healing/.

Living Landscapes Climate Science Project. An online educational project on climate issues and Indigenous peoples developed under the aegis of Salish Kootenai College and the Confederated Salish and Kootenai Tribes. Principal investigator: Adrian Leighton. Co-Investigators: Germaine White & David Rockwell. http://www.skclivinglandscapes.org.

Mission Mountains Tribal Wilderness: A Case Study. 2005. CSKT, for the Native Lands and Wilderness Council. https://www.umt.edu/media/wilderness/toolboxes/documents/IFST/

Montana Native News Project. nativenews.jour.umt.edu.

"The National Bison Range." Editorial. September 3, 2003. New York Times. https://www.nytimes.com/2003/09/03/opinion/the-national-bison-range.html.

Northern Cheyenne Tribe. www.cheyennenation.com.

Origin: A Genetic History of the Americas. 2022. Jennifer Raff.

The Place of the Falling Waters. 1991. Film. Roy Bigcrane and Thompson Smith. https://www.youtube.com/watch?v=tq6_UoLVErw.

"Program Plants, Collects Seeds of Success on Fork Belknap Reservation." July 17, 2023. https://nativephilanthropy.candid.org/news/program-plants-collects-seeds-of-success-on-fort-belknap-reservation/.

The Salish People and the Lewis and Clark Expedition. 2005; rev. ed., 2018. Salish-Pend d'Oreille Culture Committee and Elders Cultural Advisory Council. Confederated Salish and Kootenai Tribes. University of Nebraska Press.

"The Three Yayas—2022 Inductees." 2022. SQCC and Montana Outdoor Hall of Fame. https://mtoutdoorhalloffame.org/2022-inductees/the-three-yayas-2022-inductees/.

Trust in the Land: New Directions in Tribal Conservation. 2011. Beth Rose Middleton. First Peoples: New Directions in Indigenous Studies Series. University of Arizona Press.

CHAPTER 15

The Helena Region

The Gates of the Rocky Mountains.

That is what Meriwether Lewis called a spectacular limestone canyon north of Helena. On July 19, 1805, the Corps of Discovery slugged their way upstream beside the Missouri River. Their feet were sore from sharp stones and the stabs of prickly pear cactus. Lewis wrote of seeing country "unknown to the civilized world," despite Indian pictographs lining the canyon walls. Sacajawea was their guide and knew this country well. Perhaps she felt Lewis's dismissal of her worth despite his dependence on her Indigenous knowledge.

Today, we see uplifting beauty in the Gates of the Mountains, but Lewis saw dread. "Every object here wears a dark and gloomy aspect. From the water's edge the cliffs seemed to rise on either side perpendicularly to the height of 1200 feet and project rocks in many places that seemed to tumble on us." This was their first exposure to the Rockies after many weeks on the Great Plains. A central goal of the Lewis and Clark Expedition was to find a riverine Northwest Passage across North America. That geographic fantasy went unfulfilled, but they unknowingly showed that the Gates of the Mountains connected ecologically rich landscapes.

The Helena and Upper Missouri region is a link between the Greater Yellowstone and the Northern Continental Divide Ecosystem including Glacier. Animal and plant migrations can happen only if corridors of habitat are protected at all elevations, especially with climate change. This region has even broader importance as a vulnerable section of the Yellowstone to Yukon (Y2Y) Conservation Initiative.

There are many unconserved gaps between Yellowstone and Canada. North of the park, Forest Service lands are not continuous. Interstate 90 is vital for transportation, but it's a barrier for nonflying wildlife. The most intact Y2Y corridor exists along the Continental Divide through Montana, with the only significant gap at Anaconda. However, this high-elevation terrain is not suitable for all species. One of the largest topographically diverse segments is in the Helena and Upper Missouri region. Private land is widespread, so voluntary conservation easements and open space programs are essential. Fittingly, Helena is home to the Montana Land Reliance; the Prickly Pear Land Trust; The Nature Conservancy; Montana Fish, Wildlife and Parks; the Montana Association of Land Trusts; the Wilderness Land Trust, and more. Helena is the state's capital city and capital of conservation.

OPPOSITE: The Sleeping Giant dreams

Gates of the Mountains

The Missouri River cuts through the Big Belt Mountains, excavating this famous canyon into 330-million-year-old limestone. The 29,000-acre Gates of the Mountains Wilderness was established in 1964 and it's graced with 50 miles of hiking trails. The Gates of the Mountains Game Preserve provides adjacent habitat for wildlife. Boat tours in the Gates of the Mountains allow visitors to explore the canyon and visit Mann Gulch, the site of the 1949 smokejumper disaster chronicled by Norman Maclean in his book *Young Men and Fire*.

Beartooth Wildlife Management Area (WMA)

For its July 4, 1970, special edition, *Life* magazine ran a major story on "Land We've Saved," with the subheading "Conservation Victories from Florida to California." Among the places featured was the Beartooth Ranch in Montana. The Missouri River and Holter Lake form its western boundary, with the Big Belt Mountains at its core. The ranch

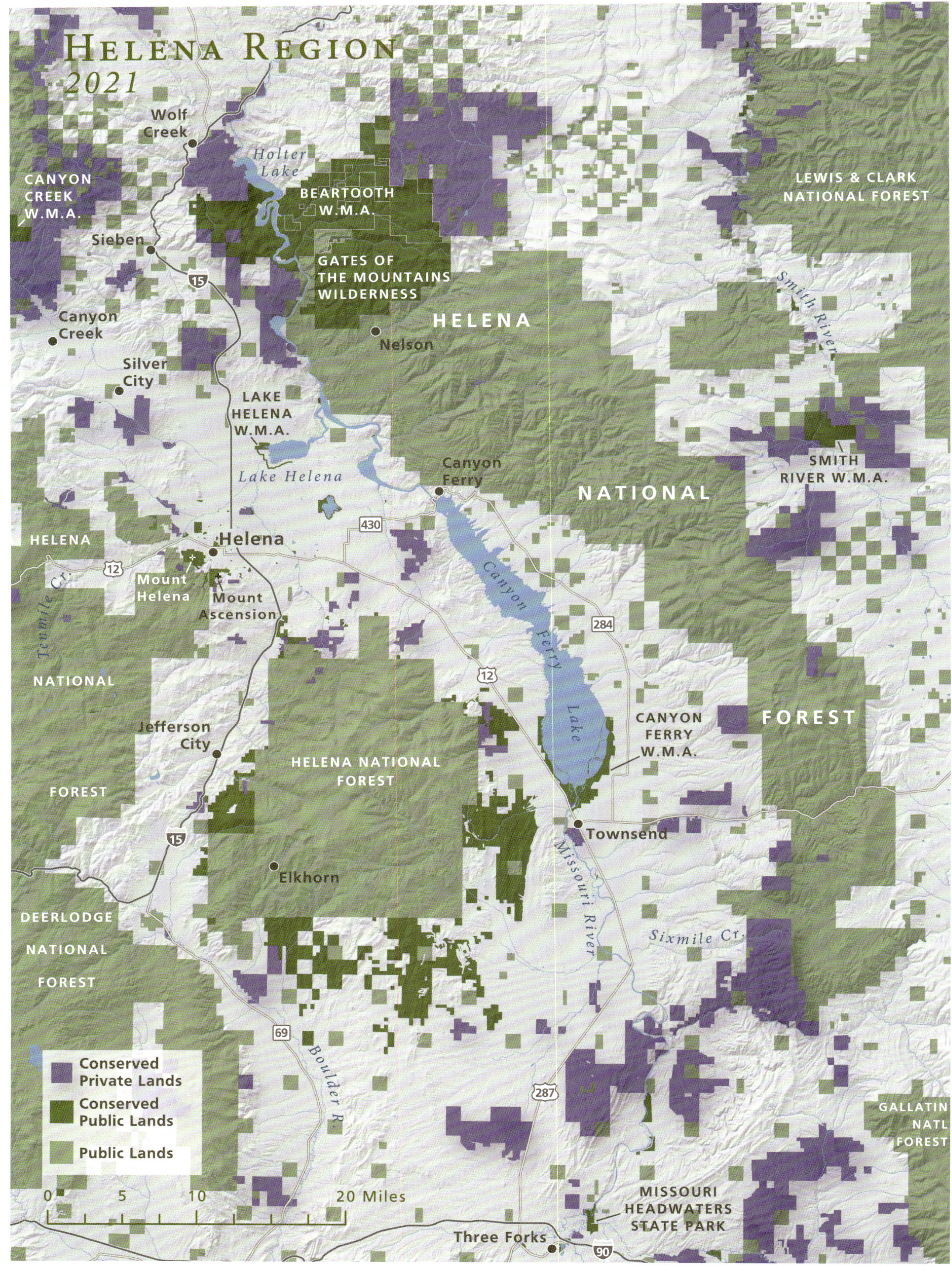

HELENA REGION
2021
Wolf Creek
Holter Lake
CANYON CREEK W.M.A.
BEARTOOTH W.M.A.
Sieben
GATES OF THE MOUNTAINS WILDERNESS
Canyon Creek
HELENA
Nelson
Silver City
LAKE HELENA W.M.A.
Lake Helena
Canyon Ferry
NATIONAL
LEWIS & CLARK NATIONAL FOREST
Smith River
SMITH RIVER W.M.A.
HELENA
Helena
Mount Helena
Mount Ascension
Tenmile Cr.
Canyon Ferry Lake
NATIONAL
FOREST
Jefferson City
HELENA NATIONAL FOREST
CANYON FERRY W.M.A.
FOREST
Townsend
Missouri River
Elkhorn
DEERLODGE NATIONAL FOREST
Sixmile Cr.
Boulder R.
Conserved Private Lands
Conserved Public Lands
Public Lands
GALLATIN NATL FOREST
0 5 10 20 Miles
MISSOURI HEADWATERS STATE PARK
Three Forks

consists of 35,000 acres of diverse plant communities and fish and wildlife habitats (especially elk and deer winter range), as well as eye-popping cliffs. The reason *Life* gave a nod to this place was that it had recently been transformed from a privately owned ranch to the Beartooth Game Range, a property owned and managed by the State of Montana. Soon, the area was renamed the Beartooth Wildlife Management Area, reflecting its status as more than just great hunting ground.

The property was previously owned by the Milton family, who operated it as a cattle ranch and private recreation area for horseback riding, hiking, camping, boating, skiing, kid outings, and neighborhood gatherings. Sadly, the ranch had to be sold because the family patriarch passed away and two other family members died in a house fire. With seven nonranching siblings as heirs, it was improbable they were ready or willing to take it on. Yet most of them agreed that conserving the ranch for future generations was a worthy goal. However, the estate executor set about the business of selling the place to a well-heeled buyer. Among the heirs, the Milton brothers were uncomfortable with that plan since they had fallen in love with a place—*this place*.

One day, 19-year-old Bill Milton wandered into a high-rise office of The Nature Conservancy in downtown San Francsico. Huey Johnson was the only staff member for a vast region that included Montana. He listened carefully as Bill spun the idea—the vision—of conserving the ranch despite the executor's determination to sell to the highest bidder. If Huey had learned anything in his short tenure, it was that dreams of saving important landscapes were not to be trifled with. He met with the executor and said TNC would like to buy the ranch at a fair price. They reached a verbal agreement sealed with an "old fashioned western handshake." However, the executor did another handshake deal with someone we're calling "Mr. New Cowboy Boots" for a higher price. Huey Johnson found out and worked out a deal where Mr. Boots would buy the place, keep the hay lands, and sell the rest of the ranch to The Nature Conservancy. It worked. TNC then transferred the land to the Montana Fish and Game Department (soon to become Montana FWP) for no cost.

This win-win transaction occurred in 1970, the beginning of this book's story arc. This solution fell within the aura of the first Earth Day and was celebrated locally and nationally. It serves as another example of a conservation victory snatched from the jaws of defeat.

Ox Bow Ranch

In 1979, former Republican governor Tim Babcock, Democratic governor Tom Judge, and Gomer Jones of the National Wildlife Federation struck a deal. The idea was to repair a 48-square-mile checkerboard of private and public land north of Helena. This included the famous Sleeping Giant rock formation and part of the Gates of the Mountains. The cooperators were Tim Babcock (who owned the Ox Bow Ranch), the Bureau of Land Management, and the State of Montana. The Ox Bow contained 12 miles of Holter Lake shoreline—land highly prized for rural subdivision. Babcock needed to be compensated for that not to occur. Over the next three years, lands were analyzed, trade-offs weighed, and a series of land exchanges and purchases worked out. As a result, a key part of the Gates of the Mountains was conserved and the Sleeping Giant Wilderness Study Area was created. It took 10 transactions and endless patience, but the parties built trust, worked out the money, and succeeded. Next time you see the belly and big nose of that reclining mountain figure, check to see if they're smiling.

Hilger Ranches

In 1865, a 20-year-old from Luxembourg named Nicholas Hilger came to Helena to mine gold. When the boom went bust in Last Chance Gulch, Hilger took up homesteading near the Gates of the Mountains. It was a summer operation back then, so "N. D." worked as a justice of the peace in Helena and later as the probate judge for Lewis and Clark County. In 1887, he bought more ground and spent $4,800 building a steel steamboat to guide tourists through the Gates. The Rose of Helena was a 55-foot-long stern-wheeler, small by Missouri River standards, but just fine to carry 60 tourists at a time. The canyon Meriwether Lewis

feared became a place for sight-seeing and picnics. A century later, the grandchildren of N. D. built up two ranches—the Hilger Hereford and the Carrie Hilger, totaling over 10,000 acres. Babe, Dan, and Bryan ran the former, but by 1983 the work became too much. In *Building the Herd*, Bryan Hilger said, "We rode into the hills after cows, one last time. We rounded up 148 head of Polled Hereford cattle and headed 'em home. Dan didn't go with us. Babe said, 'His horse died a few years back and he hasn't liked to ride since.'" The newspapers called it "The Last Round-Up." The Hilgers decided to sell, and eventually Carrie Hilger felt the same. Bryan said, "People were coming out of the woodwork, long-long, shirt-tail relatives showin' up, letters from interested buyers, real estate agents, everybody wanted to buy a Hilger ranch. We had quite a few offers, mostly from developers wanting to subdivide. But our neighbor liked the idea of a conservation easement. They promised to keep the property a working ranch, so we sold to them."

By 1985, the Montana Land Reliance had received easements on both Hilger places. The group honored the families for their stewardship. Chase Hibbard, a rancher and banker, gave them each a painting of their ranches and the Sleeping Giant they border. Chase also conserved his land nearby on the large Sieben Ranch. Babe, Dan, and Bryan lived out their days on the Hilger Hereford spread, where they are now buried. Cattle still graze and horses are still used to round up stock. Elk drop down in the winter to feed on prairie grasses. Bald eagles work the Missouri for fish. The Hilgers came to Montana, worked hard, and left a legacy worthy of the name. Their ranches are now critical parts of a conserved landscape of more than 100,000 acres. Much obliged.

Montana Fish, Wildlife and Parks

This Helena-based state agency has played a vital role in conserving wildlife habitat across the state. Its mission is to "provide for the stewardship" of Nature "while contributing to the quality of life for present and future generations." Until the early 1980s, the agency only purchased land for Wildlife Management Areas (WMAs) in places such as the Blackfoot-Clearwater region and Rocky Mountain Front. In 1985, FWP's management team learned the power of purchasing conservation easements and added this voluntary tool to its work. The Habitat Montana program was born, funded mostly by nonresident hunting license fees. Thanks to the work of Ron Marcoux, Ron Holliday, Hugh Zackheim, Debbie Dils, Darlene Edge, Jim Posewitz, Kevin League, and others, the agency holds 60 conservation easements totaling over 300,000 acres of ecologically significant private land. Much of this is big game winter range, sagebrush grasslands, and riparian habitats. Those projects are shown throughout this book.

The agency also engages in a mix of other conservation efforts focused on habitat management. These include the Migratory Bird Wetland Program (633 wetlands), the Upland Game Bird Enhancement Program, and Block Management to ensure responsible access by paying landowners who allow hunting. All this work is intended to enhance land, water, and wildlife. Hunting and fishing opportunities are essential, as are cooperating with agriculture and safeguarding open space beauty. FWP is a classic Montana success story based on respecting private property rights and rewarding land stewardship.

Road to the Gates

Prickly Pear Land Trust

Mount Ascension rises south of Helena, creating the city's natural skyline. Pine forests and prairie grasslands grow in a flower-filled mosaic on this limestone peak. Mount Ascension provides both a scenic view and a stage for the city. "Horizon, Frame, Ridge, whatever you call it, this mountain provides one of the most beautiful backdrops of any city in the United States. It is also home to a beloved trail system," notes the website of the Prickly Pear Land Trust (PPLT). Its Mount Ascension land conservation and trails project was the foundation for the group's birth and success as a regional land trust. However, back in 2003, land subdivision threatened to replace mountain habitats with houses. People gathered to create a park, and PPLT was created to fulfill that vision. After 16 years of effort involving many partners, the Mount Ascension Park was established by purchasing hundreds of acres and providing 90 miles of hiking trails in the south hills. The City of Helena manages the area. A $10 million Lewis and Clark County open space bond helped finance the deals.

Mary Hollow is the executive director of PPLT. She was born and raised in Helena, someone who knows land and life here. Mary hunts and fishes around the region. "But I was drawn to ag land conservation and trails as well as wildlife protection. We have to tie-in to communities and people. Without that, all the elk habitat protection in the world won't last."

The Prickly Pear Greenway project demonstrates the group's "Community Conservation" perspective. Prickly Pear Creek flows north through Montana City to East Helena. Former smelting operations severely degraded water quality, habitat integrity, and community health. Mary Hollow and PPLT envisioned creating a meandering open space system along the restored creek. Local communities, public schools, agencies, and other NGOs got involved. An EPA Superfund settlement repaired much of the damage. PPLT acquired 323 acres, including 10 miles of creek frontage. Land was then conveyed to communities where ADA-compliant trails were built and fishing access sites created. Schoolkids now play next to the water and

walk between home and school. A once-toxic site is now a safe haven for children and other precious wildlife species.

Fort Harrison is both an Army base and a medical center for Montana's veterans. PPLT's "Peaks to Creeks" effort protects land along Sevenmile and Tenmile Creeks in the Helena Valley. Significant funding came from the Department of Defense, as well the Lewis and Clark Open Lands Program. Military contributions created an open space buffer around Fort Harrison and the flight path to the Limestone Hills Training Area at the base of the Elkhorn Mountains. The Department of Defense understands that land development negatively impacts Army operations, and $30 million has been spent so far under the Readiness and Environmental Protection Integration (REPI) program. PPLT has been the recipient, trusted by the brass to get things done. Mary Hollow is enthusiastic about the relationship. "Military support has been wonderful. We have a strong and deep partnership with the Army." As a result, 600 acres were purchased around Fort Harrison and another 200 acres were bought to create Tenmile Creek Park. Trails and veterans' services are now available side by side. Montana is second only to Alaska in the percentage of veterans per capita. Nature offers tremendous physical and personal benefits for vets and all visitors to this quiet, conserved part of the city.

The Prickly Pear Land Trust operates in four counties: Lewis and Clark, Broadwater, Jefferson, and Powell. Its Community Conservation Program emphasizes tangible benefits for all Montanans. Mary believes that "close access to Nature improves the quality of our lives, but not every person has that. This is why we use parks, trails, as well as rural conservation easements. Parks level the playing field for folks." PPLT mirrors two other regional NGOs in this way—the Gallatin Valley Land Trust (Bozeman) and the Five Valleys Land Trust (Missoula). In contrast, The Nature Conservancy, Montana Land Reliance, Rocky Mountain Elk Foundation, Vital Ground, and Bitter Root Land Trust emphasize protecting large blocks of wildlife habitat and agricultural land. Both approaches are vital in Saving the Big Sky.

However, PPLT also negotiates conservation easements in ag country. The Potter Ranch covers 3,200 acres near Canyon Ferry Lake. It's been in the family since 1900, and Doug and Ronda Potter chose to keep it in cattle production. Doug speaks emotionally about the decision. "Our family built this up and we want to keep it in one piece, in perpetuity. It would mean a lot to my grandfather and uncles." The Canyon Cattle Company

Conservation easement on a 4.5-mile stretch of the Missouri River

Easement on the flanks of the Elkhorn Mountains

easement now protects 3,900 acres and a 4.5-mile stretch of the Missouri River from Craig to the Dearborn confluence. Funded by Lewis and Clark County open space dollars, the project ensures that fishing is available below the high-water mark and protects scenic vistas at three extremely popular fishing access sites: Spite Hill, Stickney Creek, and the Dearborn River. This reach of the Missouri generates $60 million from 150,000 anglers each year.

PPLT has helped conserve over 11,500 acres. Mary Hollow believes that "people love the Prickly Pear Land Trust because we are relevant in their lives. Republican or Democrat, it doesn't matter. Community good will get us through political divisions, all of it based on conserving the good earth we all share. We're Mom and apple pie."

Missouri-Madison River Fund

"Mitigate" means to ease pain, to soften the effect of something that may be necessary but has consequences. Hydropower dams generate needed electricity, reduce flooding, and ensure reliable irrigation water. The carbon footprint of dams is very low compared to that of power plants run on fossil fuels. However, reservoirs inundate riparian ecosystems, take ag land out of production, and alter fisheries. There are trade-offs.

The Missouri-Madison Corridor encompasses 650,000 acres of land and water—about a thousand square miles. Six privately owned dams have been built on the rivers, creating Hebgen Lake, Ennis Lake, Canyon Ferry Lake, Hauser Lake, Lake Helena, and Holter Lake. Two publicly built dams were added at Toston and Canyon Ferry. When privately owned dams are relicensed by the federal government, their negative impacts must be mitigated in some way. Montana once again provides an example of private-public partnerships for working that out.

Stakeholders in the corridor include the public, agricultural groups, NGOs, four national forests, three BLM districts, three state park districts, six counties, three cities, and many small communities. In 1992, the Montana Power Company (now NorthWestern Energy) signed a memorandum of understanding to address stakeholder concerns. High on the list was public river recreation access and water and fisheries enhancement. At the time, there were 160 developed access sites. NorthWestern Energy agreed to create 32 more, and the Missouri-Madison River Fund was born. NorthWestern Energy and public partners created a $5 million account, with interest available to be spent annually. Public recreation access managers apply for these grants, which if matched by federal and local funds are also matched by NorthWestern Energy. It's basically a two-for-one deal. Since 2007, about $12.9 million has been spent through grants to applicants for 169 projects to soften the impacts of needed hydropower facilities.

Three Forks

The Missouri River is born at Three Forks, where the Madison, Jefferson, and Gallatin Rivers meet. It accounts for nearly half of the annual flow of the Mississippi at St. Louis, as much as 70 percent during droughts. Given its hydrological size and cultural importance, the Missouri should really be called the Montana River.

The valley upstream of Helena is ranch country—wide, productive, and ecologically rich. This landscape is home to agricultural traditions that go back generations. As Helena and Bozeman grow, housing development is transforming some ranches into rural "suburbs." Landowners face the usual dilemma—cows or houses? Over the years, the Montana Land Reliance built close relationships with ag operators and now holds conservation easements on over 40,000 acres in the upper Missouri country south of Townsend. More projects are likely, given the "word of mouth" nature of life here. The focus is on family legacies, but these easements link Yellowstone with the Yukon in a place where that connection is easily broken.

At Three Forks, the Missouri emerges from a maze of channels, floodplain sloughs, oxbow lakes, subirrigated pastures, and irrigated fields—places for wildlife to find refuge and vital migration corridors. The Missouri Headwaters State Park covers 532 acres of this riparian ecosystem. You can hike, camp, fish, and soak in some of the finest water in the world—born in Yellowstone Park and still cold during summer. Rivers are

Moon on the Missouri River

restless, cutting, filling, braiding their channels, producing fresh habitat for fish, birds, mammals, hopping frogs, and slithering snakes. Look at an image on Google Maps; the confluence forms a rough triangle where the Madison, Gallatin, and Jefferson commit to each other and form a bigger flow. North of here, the Missouri enters a canyon with the Trident Cement Plant on its right bank. The prairies on either side are bone dry just a few yards from the water. After 10 miles, the valley broadens into an irrigated paradise, but then come the dams and reservoirs and our need to mitigate what has been lost. The circle re-forms in the Gates of the Mountains where land conservation offers a real prospect of eternity.

The river teaches as much as the shore.

FURTHER READING

Building the Herd. 1989. Wendy Woolett and Bryan Hilger. Tailgate Studio.

Lewis and Clark County Open Lands Program. www.lccountymt.gov.

Life magazine. July 4, 1970 (special issue).

Missouri Headwaters State Park. https://fwp.mt.gov/missouri-headwaters.

Montana Land Reliance. www.mtlandreliance.org.

Prickly Pear Land Trust. www.pricklypearlt.org.

Readiness and Environmental Protection Integration (REPI). US Department of Defense. www.repi.mil.

The Spine of the Continent: The Race to Save America's Last Best Wilderness. 2012. Mary Ellen Hannibal. Lyons Press.

CHAPTER 16

The Prairie

A squall blows in and prairie grasses bend to survive, adapting to change. The land rolls to the horizon and the sky rises forever. The heavens fetch blue vistas and 60-mile storms where lightning bolts rain fire and water onto the land. The Great Plains were once a 400-million-acre sea of big bluestem, muhly, green needlegrass, needle-and-thread, grama, and buffalo grass. If you stood on a bluff in 1800, you'd see thousands of bison, elk, mule deer, and pronghorn in one view. You'd watch gray wolves make kills and grizzly bears charge in to claim carcasses. Chuckle as the eternal coyote outsmarts them all. The Great Plains formed a shifting stage of wild animals, boundless in number and reverent in beauty. Native people lived within Nature, hunting for food and clothing, singing prayers of thanks for the Earth's many gifts. There was no sign of a beginning and no prospect of an end.

But with settlement, that balance ended as Indians were pushed onto reservations and exiled from their homelands. Free-ranging bison, grizzlies, and wolves disappeared from most of the plains, and the number of wild creatures became a fraction of the millions that once lived here. Fortunately, some of that treasure remains, sweeping across North Central Montana in rolling waves of prairie. Native Americans retain parts of it on the Fort Belknap and Fort Peck Reservations. The rest is ranched, farmed, or managed as various kinds of public land.

The Great Plains are disorienting to the untrained eye. There is no high peak as a destination, no dominant landmark; it's all the center. As Black Elk (Oglala) wrote, "The Holy Land is everywhere." Trees are sparse except in the breaks, and most of the grassland's biomass is below ground where roots build soil, store carbon, seek minerals, and draw moisture. But life is everywhere. Pronghorn gallop, swift foxes stare, and songbirds remember old harmonies. Mule deer bound away at the sight of us. Sagebrush fills the air with an aromatic musk—"cowboy cologne," some call it. Wildflowers rise in unplanted gardens of blue penstemon, orange paintbrush, and pink prairie-smoke bowing to the breeze. Sage grouse strut at their leks, jousting, cooing, booming. Elk find sanctuary in coulees. And down where the land breaks toward the Missouri River is a herd of bison. Huge, brown, and belonging. This sight was commonplace for millennia, became a memory, and is now an actual miracle.

The goal of the American Prairie nonprofit group is to create a 3.2-million-acre reserve for bison and other Great Plains species in Montana—a kind of American Serengeti, in the words of historian Dan Flores. In the Maasai language, *serengeti* means "endless plains," and the name fits here. The American Prairie is more than halfway to its acreage goal. It has built a herd of 800 bison with a target of 5,000. These remarkable achievements attract international praise as the embodiment of large landscape conservation. Some ranchers and politicians disagree, based on ideology and suspicion. Yet the effort is an exercise in private property rights and free market capitalism linked to conservation biology. In that duality is the essence of a disagreement.

Two varying White visions of land stewardship exist out here—agriculture and the restoration of Nature's Montana; feeding the world and making a home for native species. Both are virtuous and both are worthy of respect. But when visions collide, emotions matter as much as economics, and rumors travel faster than reality. Change causes excitement or fear, so facts must be sorted out.

First, let's pay homage to grasslands as the hearth of humanity. Our kind first stood upright in the plains of East Africa about two million years ago. We evolved *with* grasses. About 200,000 years ago we became fully us—*Homo sapiens*—on those same prairies and savannas. We hunted, used tools, and learned how to team up. Our brains grew and we talked to each other. We walked, then

Greater sage-grouse

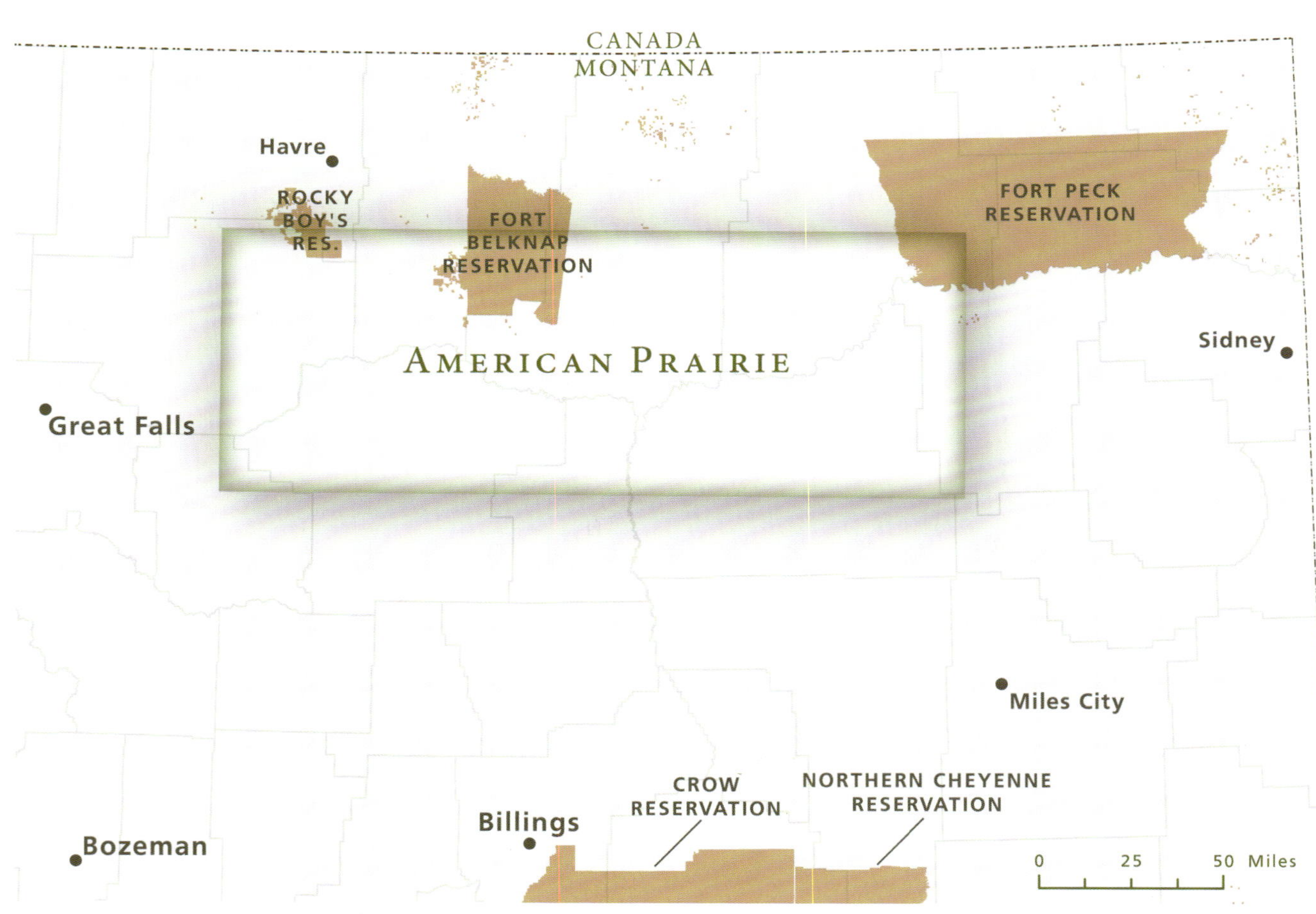

ran over open ground to hunt and gather what we needed to survive. In *American Serengeti* Dan Flores speaks about our relationship to grasslands: "It's a kind of genetic memory, because we evolved in this setting, we are entranced when we encounter it elsewhere, even though we don't know actually why." Humans then spread out of Africa, and about 10,000 years ago we domesticated cattle, horses, sheep, goats, pigs, and wheat—always depending on grasslands for survival.

This epic story began in places like Sibiloi National Park in Kenya, where human fossils back up the tale. This landscape bears some resemblance to the Great Plains of Montana. It has a dry climate and hot summers, with water in Lake Turkana—a long, slender oasis. The park has migratory waterfowl, egrets, raptors, and fish. There are large predators and large grazers, working out a balance. Cattle growers live next door, locals trying to adapt to the official preserve. Hot-button issues are grazing rights, drought, market prices, tourism, and compensation for conservation. Corridors for wildlife movement are created by paying landowners. Payments are made for livestock lost to predation.

Montana and Kenya are a world away, but the issues are similar. Where we began in Africa is where we are today in this part of Big Sky Country, trying to find a way for both ranching and a biological reserve to coexist.

A Bold Vision

George Catlin produced well-known canvases of life on the Great Plains in the mid-1800s. He painted Indian portraits, group dances, range fires, and bison rendered in vivid shades of red, blue, green, and brown. In one, a bull is surrounded by wolves; in another, Indians on horseback hunt bison, trails of blood staining the prairie. Catlin loved the landscape and respected the people. In 1841, he described a vision of being "lifted up upon an imaginary pair of wings, (and below me) was a vast and almost boundless plains of grass, which were speckled with bands of grazing buffaloes!" He proposed that part of the prairie should be "preserved in its beauty and wildness, in a magnificent park, where the world could see for ages to come . . . the fleeting herds of elks and buffaloes . . . a park containing man and beast, in all the freshness of nature's beauty." Catlin envisioned a "Nation's Park" where Native people could live beside wildlife as they always had, free of invasion and violence.

The part of Montana Catlin alluded to is home to the Assiniboine, Blackfeet, Chippewa Cree, Gros Ventre, and Sioux Nations. This is their world, so Catlin's idea may have only made sense to them later, when the cavalry, settlers, and railroads arrived. Perhaps it would have gained merit for defensive purposes, but the forces of Manifest Destiny would hear none of it. Besides, Yellowstone National Park, the world's first, was still decades away. Catlin's vision was empathetic but had no chance of taking form.

In 1914, Charles M. Russell painted a scene with thousands of bison crossing the Missouri River at sunset. A bull stares at you, face half lit in reddish brown, the other half in shadow. Wildlife bones cover the ground as skittish wolves size up their odds. Charlie called this painting *When the Land Belonged to God*. He imagined it long after bison were nearly hunted to extinction. No Indians are shown, but two years later, Charlie helped Chippewa people receive the Rocky Boy's Reservation. In 1969, the Charles M. Russell National Wildlife Refuge was established in his honor. The prairie is a gallery of dreams.

In 1987, Deborah and Frank Popper advocated creating a 90-million-acre "Buffalo Commons" in portions of 10 states. That's the size of 40 Yellowstones. The Poppers' plan was academic and out of touch with Great Plains cultures. Such a preserve would require imposing brutal federal regulations, including removing cattle from grazing leases on vast areas of public land. The Poppers saw population loss as positive and never grasped private property rights. To most residents of the Great Plains, the idea of a Buffalo Commons was high-sounding and low-down. In 1992, a Montanan named Bob Scott refined and refocused the notion on five million acres in Eastern Montana he called "The Big Open." His vision was more practical and strategic, but it never happened. By 1999, The Nature Conservancy had recommended a critical area for protecting biodiversity out on the plains adjacent to the Missouri River. It was a scientific assessment, not a plan. In 2001, the World Wildlife Fund (WWF) identified the shortgrass prairies of Montana as one of only

Missouri River cottonwood bottom

four large temperate grasslands remaining in the world. The others were in Nebraska, Mongolia, and Argentina. WWF employees Curtis Freese, Steve Forrest, and Bob Irvine, along with Tom France of the National Wildlife Federation, met with other conservationists to discuss what to do. They decided that efforts would focus on the Charles M. Russell National Wildlife Refuge, the Upper Missouri Breaks National Monument, and adjacent portions of seven counties. Interest ran high, but no one agreed to take the lead. The job was too daunting and Native people were not yet involved.

The American Prairie

In 2002, Sean and Kayla Gerrity took on the challenge with no money in the bank and no certain path ahead. Curtis Freese, a biologist, was a cofounder and ecological guide. Their audacious calling was to create a 3.2-million-acre nature reserve for bison where previous efforts had stalled. In his book *Back from Collapse: American Prairie and the Restoration of Great Plains Wildlife,* Freese wrote that this 5,000-square-mile expanse was big enough to sustain viable populations of diverse wildlife species. So did Dale Lott in *American Bison: A Natural History.* Sean and Kayla were both graduates of Montana State University with degrees in psychology, not natural sciences. Hardly the background of most conservationists, but one that eventually brought huge benefits. Back then, the couple ran Catalyst Consulting in Santa Cruz, California, a firm focused on industrial psychology, helping Silicon Valley nerds learn how to work together. By the late 1990s, with business booming and staff growing, the Gerritys felt a strong desire to return to Montana to raise a family and enjoy the outdoors. Kayla, a regular kid from Helena, said that "Montana shaped me by being outside so much."

Sean was from Great Falls, where his dad swept floors at Malmstrom Air Force Base. Sean and his family lived in a single-wide trailer and spent nearly every weekend in the boonies, where "I learned to track, hunt, and fish."

In 2004, this couple and Curtis Freese formed the American Prairie Foundation as "our next big challenge." Later, it became known as the American Prairie. They had no illusions. "Big things are always hard," Sean Gerrity said. "You build from nothing to something, and bust your butt for years. Our idea was to create a free-standing organization rather than wrangle a lot of NGOs toward the task." Kayla set up operations and management systems. "It felt like a whirlwind," she said. Sean agreed to serve as executive director for only six months, "but then it all got out of hand" and he stayed on.

Curtis Freese worked as the director of WWF's Northern Great Plains Program. He believed that this ecosystem had been viewed as "Conservation Flyover Country" for far too long. That changed when WWF designated that region as one of its global priority ecoregions. The group kicked in some money and in-kind help to get things rolling, but it wasn't nearly enough to make real progress. Sean Gerrity and Curt Freese took a gamble and pitched the American Prairie idea to the Packard Foundation. They brought a highway map to the meeting with an orange line sketched across North Central Montana. After the pitch, Susan Packard stood up and said, "I've never heard of anything so exciting in my life!" Two months later, the American Prairie Foundation received a check for $80,000. Nice, but it needed $500 million to accomplish its goal. There was every chance the whole deal would fall through, but with the help of donors, it now thrives. The idea is too grand to refuse.

The idea takes shape

American Prairie Land Projects

Assembling property is the primary task of the reserve, stitching together purchases, leases, and existing federal lands to provide wildlife habitat. The Great Plains of Montana were chosen over the Sand Hills of Nebraska because of that promising mix of land tenure. The 3.2-million-acre target is seen as "enough to support a healthy prairie ecosystem with 5,000 bison." To date the group has identified or assembled over 1.9 million acres, 60 percent of the land needed. This includes

- the 1.1-million-acre Charles M. Russell National Wildlife Refuge
- the 377,000-acre Upper Missouri Breaks National Monument
- 335,000 acres of leased public land
- 121,000 acres owned by the American Prairie

Thirty-six local ranches have been voluntarily sold to the foundation. These "base properties" are attached to leases with the Bureau of Land Management and the State of Montana. More projects are added each year. The American Prairie is not a monolithic expanse; it consists of 10 units along the Missouri from the PN Ranch in the west to Timber Creek in the east. Blocks of deeded land (with attached grazing leases) are found near the Charles M. Russell National Wildlife Refuge and the Upper Missouri Breaks National Monument. All of this is intermingled with private ag lands and public leases not held by the Prairie. Locals still control most land in the region and use it as they see fit. The Fort Belknap and Rocky Boy's Reservations are unattached to the reserve but are bonded to the project because of bison. More on that soon.

The American Prairie operates in seven counties: Blaine, Choteau, Fergus, Garfield, Petroleum, Phillips, and Valley. These rural places are dominated by cattle grazing and wheat farming. Local people are politically conservative and suspicious of change coming in from the outside, especially change that takes land out of production. Life is rewarding but difficult in this remote area, and it's been that way a long time. When the open range ended in the late 1880s, ranching required expensive fencing, water development, and predator control.

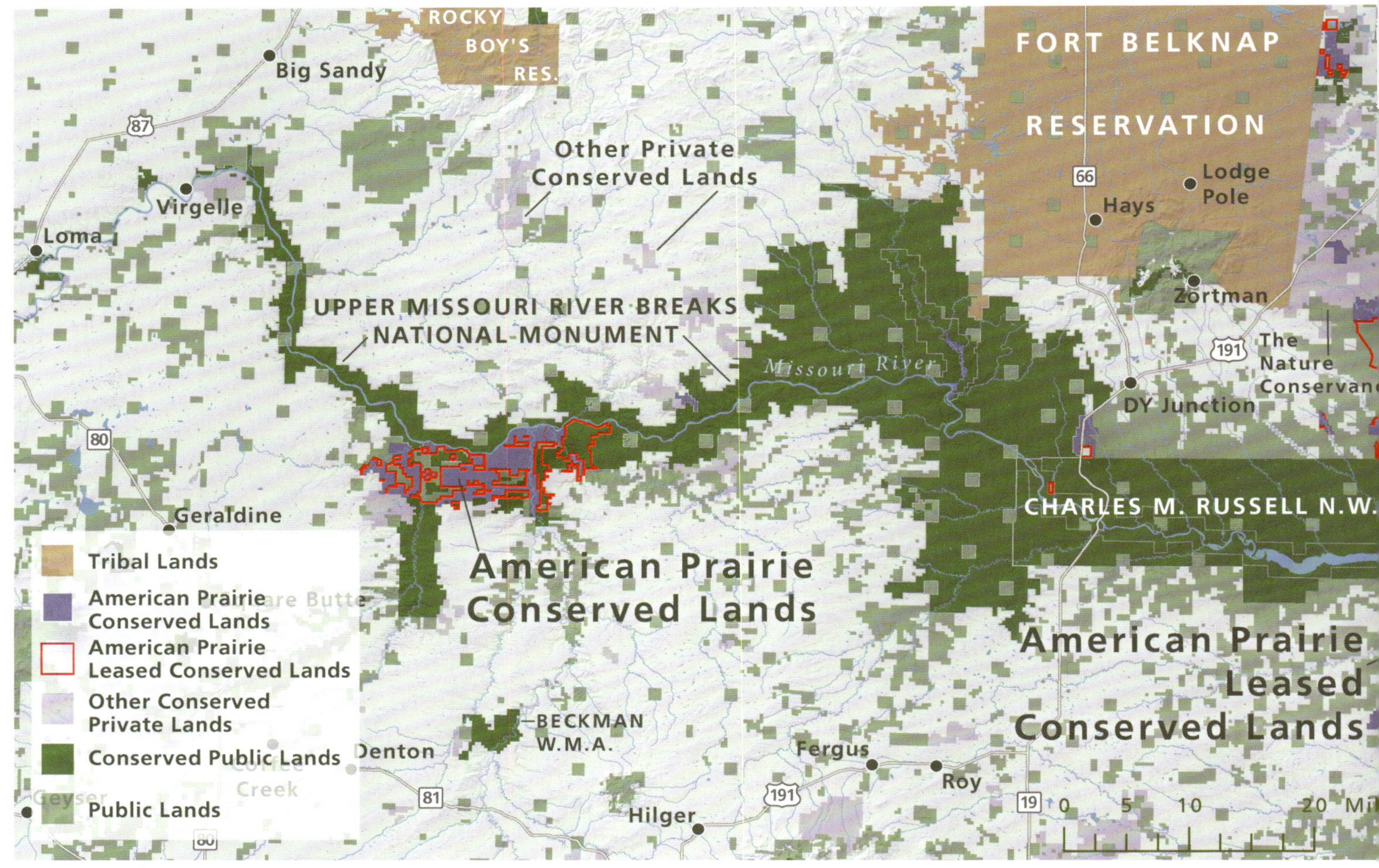

Many homesteads eventually went under because they were too small, too dry, and too unproductive. Those properties were reassembled as ranches or reverted to federal ownership. Grazing leases were hammered out, but these can feel impermanent to ranchers—a cause of real worry because the economy has often let operators down. In the past 50 years, a rancher's share of every dollar spent on beef has declined from 60 cents to 39 cents. There are fewer meat-packing plants in Montana, and prices get fixed to the detriment of producers. Large areas in the region are planted in spring and winter wheat. The price of a bushel has risen from $1.65 to $8.50 in the past 50 years, but drought and volatile world markets make farming a challenge. Inflation significantly raises operating expenses and cuts profit margins. The children of ag operators sometimes leave rather than continue the struggle. Then more drought comes and more people move out. Since 1920, the seven-county population has dropped by more than half, from 77,000 to 38,000.

The region remains overwhelmingly agricultural. There are about 464,000 head of cattle and just 800 bison. Of 1,813 large ag properties (1,000 acres or more), the Prairie owns 36 (2 percent). Of 10.6 million acres of private land in the seven counties, the Prairie owns 121,000 (1 percent). If the 3.2-million-acre goal is achieved, the Charles M. Russell National Wildlife Refuge and the Upper Missouri Breaks National Monument would stay the same. So, here's what the American Prairie might control at "completion":

- 500,000 acres owned
- 1,225,000 acres leased
- 1,725,000 total acres under Prairie management (8 percent of land in the seven counties)

Eight percent is a significant figure, but Sean Gerrity says, "We're about twenty years away from getting this done. We are optimistic, but time will tell what the final mix of land turns out to be."

Those are the basic facts about land assembly. The Prairie is a major player, but even if its plan comes to fruition, 92 percent of the seven-county area will remain as it is. This doesn't stop the worry among some neighbors, and a few are downright ticked off. Local billboards read, "Don't Buffalo Me: No Federal Land Grab" and "Save

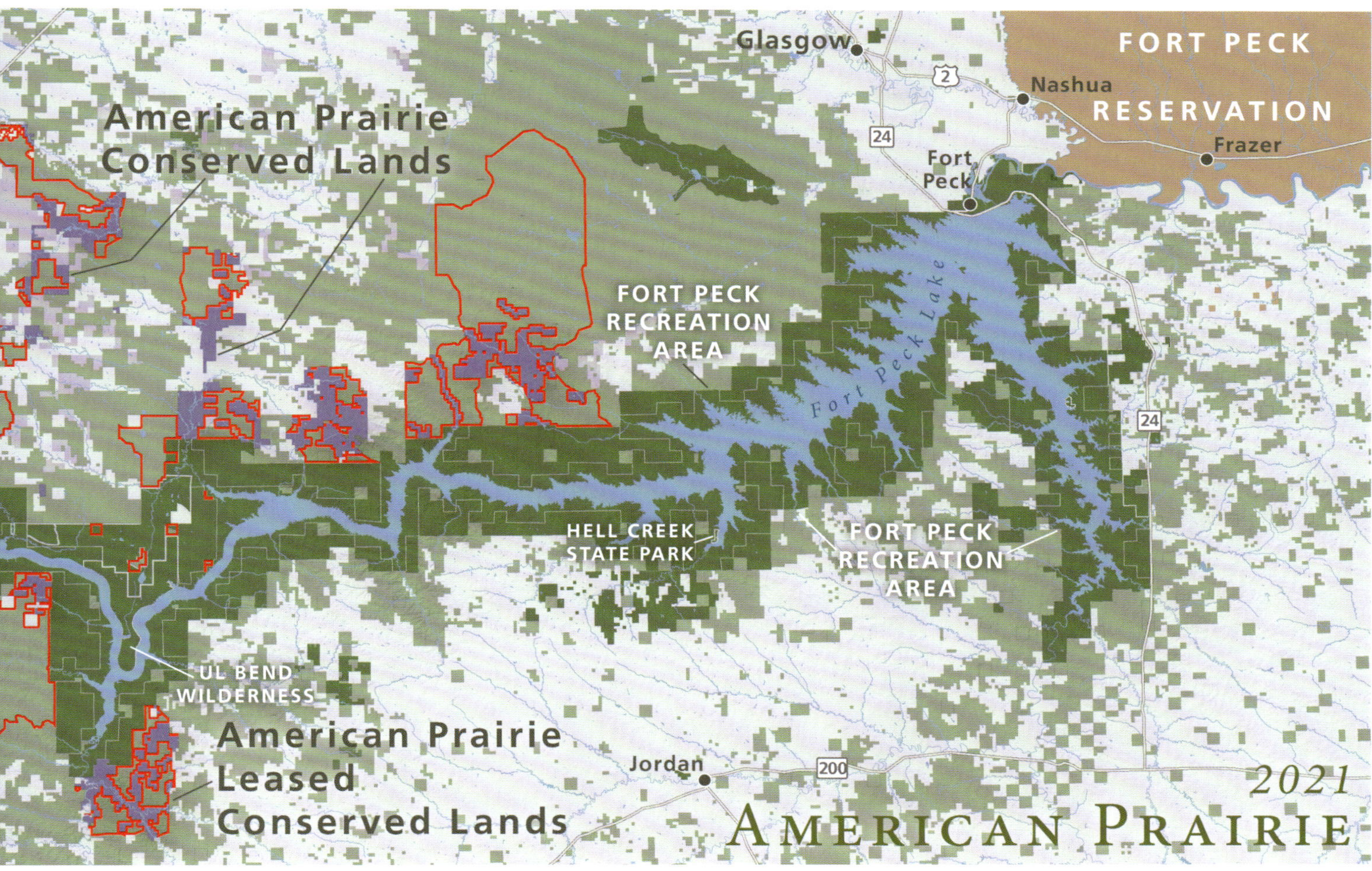

the Cowboy, Stop American Prairie Reserve." A rancher wrote to the *Billings Gazette*, "If the American Prairie succeeds, local farms and ranches have no choice but to sell." Another says, "They're not only destroying our way of life, they're destroying a vital economic base which is the foundation of America." The rhetoric gets overblown, but it cannot be ignored. Some concerns are genuine.

Restoration

The Prairie seeks "to fully restore the shortgrass prairie ecosystem on the Montana plains." Its reference point is the pre-White condition in about 1800—what Native people knew and Lewis and Clark tried to capture in their journals. There are no scientific studies from that time, only impressions and oral histories. One morning on the prairie Lewis wrote, "We have an extensive and most enchanting view . . . in every direction was one vast plain in which unnumberable herds of Buffalow were seen attended by their shepperds, the wolves." They marveled at great blue herons, elk, and pronghorn. On May 20, 1805, Lewis wrote in his journal where the Musselshell River joins the Missouri: "This broken country pretty well-timbered particularly at its borders, and interspersed with handsome fertile plains and grasslands."

The journals of Lewis and Clark serve as a rough ecological baseline study in the heart of today's American Prairie. The landscape seemed pristine to them because of the stewardship of Native American people, but the Prairie's restoration is an ongoing experiment. What *was* the land like 200 years ago? Pack rats may help sort this out. Lewis and Clark saw bushy-tailed wood rats (pack rats) on slopes above the Missouri River and included a drawing in their journals. Pack rats harvest plants and deposit them in caves and under rocky overhangs to build middens. Those piles of leaves, stems, branches, and seeds are a faithful ledger of whatever grew within 200 feet of a pack rat's home. These talented rodents pee on the pile, and what isn't eaten gets fossilized in records going back 12,000 years. Radiocarbon dating of layers in middens is a reliable way of understanding ecology through time. Basically, while Lewis

Red fox

and Clark recorded the presence of pack rats, rodents recorded prairie conditions in 1804 and 1805. Their middens show that the landscape was a mosaic of mixed-grass prairies and sagebrush steppes with conifers in the breaks, much as it is today in relatively intact places.

Restoring the prairie is a challenging job. What works? How do you measure "success"? Are 5,000 square miles and 5,000 bison sufficient? These questions are data driven, emotional, and political. The American Prairie also wants to return or expand populations of beavers, black-footed ferrets, black-tailed prairie dogs, sage grouse, ferruginous hawks, snowy owls, pronghorn, mountain lions, elk, mule deer, black bears, and eventually wolves and grizzly bears. Each species is a novel. For conservationists it sounds magical, but for some ranchers it sounds like a plot. Certainly not elk and deer, but a return of predators can be confounding.

The American Prairie website explains it this way: "Biodiversity builds resiliency. Research has shown that areas with high amounts of native biodiversity, including plants, insects, and animals, are more likely to endure the harsh conditions of climate change." Its restoration goal is also psychological: "We need to stretch our legs and minds. Actively spending time in nature is good for our bodies, brains, and emotional well-being, and the public should be able to enjoy the land with ease and without fences or 'No Trespassing' signs." The organization stresses "Rewilding"—a return of what was lost. The disparity of views between these statements and some ranchers is clear. However, beyond land acquisitions and vegetation recovery, a major conflict concerns the resurrection of a nearly vanished creature.

Bison

Bison are the signature species of the American Prairie. They are both a charismatic symbol and the connective tissue that links ecology, ecotourism, and Native American cultures. Bison stir

our imagination and rile our grief. In the 1880s, the last wild herd of bison in the Great Plains was found between the Bears Paw and Little Rocky Mountains. This landscape is in the shared homeland of the White Clay People (Gros Ventre) and the tribe called One Who Cooks with Stones (Assiniboine).

In 1800, there may have been 60 million bison in North America. By 1889, there were only 1,000. Market hunting, habitat loss to settlement, and wars against Indians were to blame. Slaughtering bison was about meat, bones, and hides—full-bore commerce by regular people—but it was also a federal instrument of war. A companion of Buffalo Bill Cody said, "Kill every buffalo you can! Every buffalo dead is an Indian gone." General Philip Sheridan understood that as long as bison herds remained, Indian people would resist conquest. He wrote, "If I could learn that every buffalo in the northern herd (Montana) was killed I would be glad. The destruction of that herd would do more to keep Indians quiet than anything else that could happen. I think it would be wise to invite all the sportsmen of England and America there for a Grand Buffalo hunt and make one grand sweep of them all." Later, Sheridan would call for an end to bison hunting because the government would have to feed Indian people, an expense he opposed. John Fire Lame Deer (Oglala) explained how brutal this time was. "The buffalo gave us everything we needed to survive. His hide was our bed, our blanket, our coat. It was our drum, throbbing through the night, holy. His horns were our spoons, the bones our knives, his sinews our bowstrings. His mighty skull was our altar. Without bison we are nothing." Native people still revere bison as an enduring spiritual being. In 2016, bison became the official national animal of the United States, taking its place beside the bald eagle as a biological ideal. Bringing this species back to the prairie is about more than natural science; it is about history and redeeming ourselves from an original sin.

After the die-off, cattle replaced bison across the plains. Resistance to the restoration of Native

Elk at home on the prairie

herds is both understandable and curious. Bison and cattle are biological cousins that share a common ancestor going back five million years, but they are now distinct in meaningful ways. The vernacular name "buffalo" suggests a connection to Cape buffalo in Africa, but bison are quite different. Bison diversified in Asia and their closest genetic relative is the yak. Cattle were bred mostly in Turkey, Iran, and Europe and then joined humans as we spread around the world. Despite all that, bison and cattle remain similar enough to crossbreed and produce the beefalo, a hybrid inferior to both.

Bison and cattle are cousins, but they are *very* different animals. Both thrive on Great Plains grasses, producing healthy, high-protein meat. Yet bison are native to this land, eat a wider range of plants, and can handle steep terrain. They also wander more, cropping grasses and moving on. Bison are migratory landscape engineers because they eat and run, letting grasslands regenerate. They also spend less time near water than cattle. Both species are vulnerable to predation, although wild bison have the advantage of size, horns, greater speed, and protective herd behaviors. Cattle tend to obey fences; bison are fond of knocking them down. Bison can deal with brutal cold far better than cows, as seen during the winter of 1886–1887 in Montana. In places, 90 percent of the cattle died, but bison were far less affected. So, should we all raise bison on ranches and not cows? Probably not.

Bison are not fully wild or fully domesticated, remaining rank and aggressive toward humans. Running bison is much more challenging and expensive than managing cattle. Beef is a staple of diets around the world, while bison is a tiny market share. So far, it doesn't pay on a large scale. Cattle ranching is well established and efficient, supported by proud and proven systems of production. It seems certain to remain the dominant land use near the American Prairie into the future. But a disease called brucellosis scares ranchers.

Brucellosis is a contagious bacterial infection that impacts both wild species and cattle. Bison and elk are known carriers, but cases of transmission to livestock are extremely rare. The disease is spread when an infected fetus is aborted and another animal licks the carcass. In 2022, two cases were found near Yellowstone National Park and both cows were killed to halt the spread. Both infections were transmitted from elk to cows. Montana averages one case per decade, and none have been found on or near the American Prairie.

Bison are classified as livestock by the State of Montana and federal agencies, which is why the American Prairie manages numerous grazing leases on public land. All bison brought to the reserve go through testing for brucellosis and are vaccinated. No cases exist to date. All bison are also genetically checked to determine their relative "purity" since most bison have some cattle DNA. A bison that acts wild is basically what the American Prairie is looking for—the opposite of what works on a ranch. Bison and cattle serve very different purposes.

On the cold, rainy night of October 20, 2005, the first 16 animals were released into the American Prairie landscape. The next spring, five baby bison were born. Their parents came from Wind Cave National Park in South Dakota and herds run by The Nature Conservancy. Five years later, another 94 bison were added from Elk Island National Park in Alberta. Each year more calves are born. Today, about 800 bison roam different parts of the American Prairie from the Judith River to Fort Peck Lake. Serious conflicts are rare. Bison sometimes cross exterior fences on snowdrifts and have to be rounded up. The American Prairie has staff on the ground all year working with neighbors.

The reserve frequently donates bison to Native American tribes across Montana so they can build their own genetically healthy herds. This totemic animal is back home on many reservations, from Fort Belknap to Rocky Boy's, Blackfeet, and Flathead. Bison hunts have returned to tribes and meat is eaten as it was before, a model for resource management linked to dignity. Native people come to the American Prairie to animate customs and engage in a spiritual return. The NGO also manages a hunt inside the reserve, receiving about 3,500 applicants and awarding 40 tags each year. Strong preference is given to local people. This new hunt brings back something old—a chance to kill a wild bison in cold weather on your own two feet. You have to walk in, take your shot, butcher the animal, and carry out the meat. Hunting stories are beginning to be told, even bragged about in local bars. The reserve is not a zoo.

Compensation

Sean Gerrity says, "We find more and more ways of providing financial and personal benefits to our neighbors." The American Prairie buys ranches using money raised from donors, *not* government programs. There is no "federal land grab." The American Prairie's annual budget is $10 million, which supports local towns when money is spent on supplies, repairs, vehicles, food, and gear. It pays full property taxes on all its land, despite no legal obligation to do so. The National Discovery Center in Lewistown is visited by tourists who stick around to spend money at local businesses. The Wildlife Friendly Lands Program pays landowners who adapt to wildlife, as much as $15,000 per year per family.

The Wild Sky Program is extremely inventive. Ranchers outside the reserve can choose to allow wildlife cameras on their property. For each sighting of certain species, they get paid; the amounts include $50 per coyote, $350 per mountain lion, $500 per black bear, and $1,000 per grizzly bear (no sightings yet). One rancher hit the jackpot when a mountain lion mom was photographed with three cubs—a $1,400 check paid on the spot. Eighteen ranches are involved in the program, with more applicants waiting than the Prairie can afford. One rancher got paid $15,000 in one year, enough to buy some new cattle or pay off some old bills. Caps and T-shirts emblazoned with Wild Sky are now seen in towns across the region.

The 60,000-acre Matador Ranch (shown as light purple on the map) is not part of the American Prairie project, but it offers a different kind of payment to ranchers called a "Grass Bank." The Nature Conservancy owns the property and welcomes ranchers to graze cattle at a discounted rate in exchange for "wildlife friendly practices on their own operations." It's a trade. Ranchers pledge to control weeds and not plow their prairies. For even lower grazing fees, ranchers agree to "protect prairie dog towns, secure Sage Grouse leks, or modify fences to make them safer for wildlife." A rancher named Dale Veseth says, "We are real pleased with this opportunity. This made

Prairie conservation: something to sing about

a huge difference in this community. When you help feed families and cows, they'll remember." The Matador Ranch conserves grassland ecosystems, but it is operated much differently than the American Prairie.

There are critics of "compensation for stewardship" efforts, but many ag operators already receive federal money through the Conservation Reserve Program (CRP). This voluntary USDA effort provides annual payments to landowners who fallow marginal cropland, seed the land to native grasses, restore wetlands, and leave buffer strips along riparian areas. This popular program was signed into law by President Ronald Reagan in 1985. Landowners sign 10- to 15-year CRP contracts and an annual payment arrives in the mailbox. According to the USDA Farm Service Agency, over $34 million was received in the seven counties between 2017 and 2021. The US Fish and Wildlife Service has also purchased perpetual conservation easements on over 450,000 acres of working cattle ranches in this part of Montana.

In many ways, conservation pays.

But compensation is about more than money. It's about the pride ranchers feel in providing for their family and feeding our country. Or the joy a visitor feels standing on a bluff, listening to a conservatory of natural music. This region offers both kinds of rewards, but people's values can collide in seemingly insolvable ways. Is the American Prairie a retreat from customs or an expansion of honor? Are the Great Plains a kingdom of cattle or an American Serengeti?

Can they be both?

Resolution

Sean Gerrity tells a story about sitting with an elderly rancher. The coffee has gone cold in their cups. A clock ticks on the wall. The rancher speaks softly, explaining how his kids left years before, leaving him to run things alone. He says they let him down, let ranching down. Times are changing and to him the old ways feel attacked. "Why don't people want what we want?" the rancher whispers. The man grieves, staring at the floor. Finally, he looks up and asks, "What happened?"

Gerrity has heard this many times. "Some ranchers genuinely feel we're wrong. Not just wrong, but disrespectful of what they do. Cattle producers often say 'We feed the world—we prevent people from starving. No one knows how hard we work. Why are you American Prairie folks taking good land out of production? It took a long time to run off the wolves, get rid of prairie dogs, and build all that fence.' I've learned that most ranchers are good people, but our work really angers some of them. They remember when 70 neighbors used to gather in a hall with a fiddler playing, kids running around, tables full of food, with whiskey shared in the parking lot. It was a golden time and must have been wonderful. But the Prairie project didn't hurt all that; economics, drought, out-migration, a dozen things did. And bison mostly get the blame."

Contrast that grief and anger with the positive feeling inside the American Prairie's office in Lewistown. The staff is mostly younger people committed to the mission. They move around energetically, upbeat and smiling, feeling they are part of creating something new and vital. Kayla Gerrity says, "We deal with the conflict when we have to, but we keep our focus on the productive work in front of us. News reports call us isolated in a place where the majority don't want us around. Few tell the story of all the good we're doing. Truth is, most people out here quietly stay out of the argument." In 2017, Alison Fox was named the new president of the American Prairie organization. She said, "I hope the theme of the coming years is inclusion. I see huge opportunities to engage a much broader audience in this vision."

There are fundamental differences between the American Prairie project and conservation easements that ranchers place on their land. Both protect wildlife habitat and scenic beauty, but the former is about keeping cows out, while easements are about keeping livestock and crop production strong. The two approaches should not be confused. The American Prairie is an exception to the usual work being done by land trusts and land-managing agencies in Montana.

In time, the American Prairie will probably settle in as part of the landscape just like Glacier National Park, which was also strongly opposed in its early years. People will agree to disagree, then set things aside. The American Prairie is only 20 years old, a newcomer as locals see it. But everyone was a newcomer once—a settler in the homelands of Native American people, a pilgrim in a realm of elk and bison. The Gospel of Luke teaches the story of the Good Samaritan, someone who helps a stranger in need. Scripture teaches that all souls seek community and respect. That we need each other. Perhaps, diverse gatherings at "cowboy church" would improve understanding. Or time spent together on the American Prairie in a cathedral of unfenced wind. Resolution arrives when we resolve to try.

Mars Vista

This beautiful part of the reserve borders the Charles M. Russell National Wildlife Refuge and the Upper Missouri Breaks National Monument. The land is a mix of flat grasslands and open stands of ponderosa pine in coulees leading down to the river. The continental ice sheet ended here, forming the edge of the Glaciated Great Plains region. But Mars Vista is not remote; it is accessible on a paved road connecting Malta and Lewistown. Semitrucks slug their way uphill from the water, and tourists poke around. The place's unusual name comes from Jacqueline Badger Mars, a conservationist who is part of the candy family—Mars Bar and Milky Way. Jacqueline is a board member of the American Prairie who generously donated the funds to buy this piece of land. It is one of 10 units on the reserve but offers a singular message.

In 2023, scientists discovered the remains of a glacier on Mars. Not ice, but sulfate salt deposits preserving the shape and features of a glacier melted away long ago. There are moraines, crevasses, and evidence of past water. That is remarkable, but the topography is staggering in its implications. This relict "glacier" stands where the land breaks down to a west-to-east channel that once held a large river. It is called Valles Marineris—the Mars Valley. Around it are mountains, plains, sand dunes, and a labyrinth of tributary channels. This is a striking terrain analog for the American Prairie, and the maps almost match up. Mars is just 125 million miles away, close enough for us to see in high definition right now, and walk on in only a few years. Turns out that what happens down here happens out there.

The real illumination at Mars Vista is the sky. Up to the red planet, a place once wet and probably home to life. Mars was born at the same time as Earth, but the two planets diverged. Mars became rusty and barren; our home became blue, green, and gold. How unlikely is the world we live in? How precious is Nature itself? Humility grows in places like Mars Vista, where our disagreements seem smaller and more easily released. Reverence and wonder come naturally here. Amazement is not the exclusive right of children.

The American Prairie is just one vision of land stewardship. There is room for many others in this vast but small corner of a blessed planet in an infinite universe. Regardless of our politics and faith, dominion is best understood as a call to wisdom in all its forms. Accepting that is not only decent; it is American.

Long-billed curlew

FURTHER READING

American Bison: A Natural History. Organisms and Environments 6. 2002. Dale Lott. University of California Press.

American Prairie. www.americanprairie.org.

American Serengeti: The Last Big Animals of the Great Plains. 2017. Dan Flores. University Press of Kansas.

Back from Collapse: American Prairie and the Restoration of Great Plains Wildlife. 2023. Curtis H. Freese. University of Nebraska Press (a detailed examination of the cultural and biological story of the American Prairie project).

Letters and Notes on the Manners, Customs, and Conditions of North American Indians. 2015. George Catlin. Palala Press.

"Prairie Divide: An Ambitious Plan to Return Grasslands of Central Montana to the Wild Splendor of the Past Faces Impassioned Resistance from the Present." February 2020. Hannah Nordhaus. *National Geographic*.

"A Relict Glacier near Mars' Equator: Evidence for Recent Glaciation and Volcanism." 54th Lunar and Planetary Science Conference (LPI Contribution No. 2806). 2023. Pascal Lee. https://www.researchgate.net/publication/369369563 (see fig. 1, relict glacier location map).

Where the Buffalo Roam: The Storm over the Revolutionary Plan to Restore America's Great Plains. 1992. Anne Matthews. Grove Weidenfeld (analyzes the Poppers' Buffalo Commons proposal).

"Wild Bison Restoration: The Suitability of Montana's Big Open." 1992. Robert Scott. *Restoration and Management Notes* 10, no. 1: 51–52.

OPPOSITE: Road to the heavens

CHAPTER 17

The Crucial Role of Capital

Across Montana, from valley to valley, region to region, you've seen how capital plays a crucial role in land conservation. Values matter, but so does valuation. Recognizing that fact makes seemingly impossible tasks feasible and achievable. Linking the constitutional protection of private property rights to capitalism brings astonishing strength to land conservation work. Even in a time of political division, this sensible approach is widely accepted.

Money works in ways that stewardship ethics alone cannot match. It is a way of tangibly thanking people for conserving their land. Financial compensation is neighborly and honest. And it works. The array of capital-funded land conservation tools is vast and creatively expanding. Even if a love of place is not the primary motivation for landowners, conservation easements offer a means of accomplishing personal and family goals by capturing unused development value.

Donated conservation easements rely on state and federal income and estate tax benefits thanks to bipartisan support in Congress. Easement and land purchases require cash, and sources are expanding every day. Land exchanges repair destructive "checkerboard" ownership patterns. The Farm Bill funds the Conservation Reserve Program to buy term easements on agricultural land—a very popular program in Montana. The Agricultural Conservation Easement Program funds the purchase of perpetual easements. City and county open space bonds allow communities to buy land for public recreation use, build ball fields, create hiking areas, and protect wildlife habitats. The Land and Water Conservation Fund transfers payments from offshore oil and gas development to land protection and recreational facilities. Dam relicensing generates funds for fishing access sites. The Department of Defense uses Readiness and Environmental Protection Integration (REPI) funds to purchase easements on land buffering military bases, ensuring our national security. The Forest Legacy Program, conservation real estate companies, conservation buyer deals, leveraged capital, philanthropy, conservation-focused banks, and more play their roles. The menu is long and splendid.

Spencer Beebe has been raising money and conserving land for more than half a century. He worked for The Nature Conservancy in the Pacific Northwest, including Montana, then around the world. He then founded Ecotrust, a Pacific Northwest–based NGO whose goal is "to foster a natural model of development that creates more resilient communities, economies, and ecosystems around the world." Beebe believes in "directing capitalism toward conservation in ways that benefit human beings by building a portfolio of healthy land and water." Spencer has a shock of white hair that reminds you of Albert Einstein. The image fits. He is a genius at inventing ways of financing the protection and wise use of key lands. Spencer says, "It's about leverage: modest amounts of philanthropy can leverage immense sums."

Spencer Beebe was one of the first people to use "Debt for Nature Swaps." We haven't seen it in Montana yet, but here's how it works. In the 1980s, he secured a $100,000 donation from the Widen Family Foundation and used it to buy Bolivian government bonds. Strange sounding, but hang in there. When the note matured, Bolivia owed his organization $850,000. Spencer tells the tale with a twinkle in his eye. "They said we don't have the money. So, I responded, OK, but I've got a deal for you. Set up a large rainforest reserve and I will tear up the bond." Bolivia then established the 400,000-acre Beni Biosphere Reserve. The landscape continues to support incredible biodiversity: 100 species of mammals, 500 species of birds, and more than 2,000 plant species. It is also the home of 215,000 Indigenous people who earn a living within the reserve through sustainable hunting, fishing, plant gathering, and traditional agriculture. The land is managed for long-term health and reliable incomes. Conservation easements in Montana also demonstrate how working lands are vital for economic and ecosystem health.

But good old-fashioned philanthropy still counts. Spencer Beebe offers a straightforward way of asking for donations. "Propose a land conservation project that makes sense. Then sit eyeball to eyeball with a person who can afford to help. Then ask directly for money. Then be quiet. Let them respond yes or no." This seems easy enough, but Spencer knows the key is building relationships with people. "Some have lots of money, but they are dying to bring meaning to their lives." The same holds true with landowners in Montana, wealthy or not, who want to leave a legacy of good land. We owe them a debt of gratitude for the millions of conserved acres across the state. Assembling this book has been humbling. Montanans are generous, straightforward people. We have mentioned so many names in this book because they deserve to be honored.

A simple truth lays the foundation of all this: landowners don't have to conserve their land if they don't want to. If it isn't the right time, or the compensation isn't enough, or they have other plans—that is their absolute right. No hard feelings, no conflict. Voluntary, market-based options rest at the core of all you have seen in *Saving the Big Sky*. As well as respect, relationships, reciprocity, and rights. Spencer Beebe shakes his head in amazement. "Conservation finance in North America—there's nothing quite like it in the world."

Story Clark agrees. She has been raising capital and conserving land in the Mountain West for decades. Story is also a landowner who placed a conservation easement on the family ranch and served on the board of the Wyoming Stock Growers Land Trust. She is enthusiastic about the work. "I spend my days researching and developing new conservation finance techniques. Then I share them." Her book *A Field Guide to Conservation Finance* sits on the desk of every practitioner in the country. Terry Tempest Williams says, "Who could imagine a book about money would be a book about love. Raising money is always about relationships toward each other and the land." The second volume of the book covers topics like conservation developments (cluster developments with easements), conservation lenders and investors, off-site mitigation, and carbon markets. Story is well worth listening to about economics. She and many others are forging new ways of creating prosperous communities based on healthy environments.

Historian Mark Milliorn says it clearly: "If you want more of something, reward it. If you want less of something, don't reward it."

Oikos means "the house." Our home, the root word of both economics and ecology. It turns out that capital is attracted to whatever we ask of it, and this revelation has transformed how conservationists work. Rather than oppose things, they propose solutions. There's elegance to this understanding, a way of moving through life that builds trust with others. Maybe it's time to amend an old saying.

Money talks, but sometimes it sings.

FURTHER READING

Cache: Creating Natural Economies. 2010. Spencer Beede. Ecotrust.

A Field Guide to Conservation Finance. 2007. Story Clark. Island Press.

A Field Guide to Conservation Finance II. 2024. Story Clark. Island Press.

Principles of Sustainable Finance. 2019. Dick Schoenmaker and William Schramade. Oxford University Press.

CHAPTER 18

Making Certain It Goes On

Over the past half century six million acres of private land have been conserved in Montana. By the time this book is printed, that figure will have grown. Our timeline began in 1970 and ended in 2021 because of map and book production schedules—a roughly 50-year arc. Rest assured that the patterns you see in this book continue to be reinforced by additional projects. Another 100,000 acres of forest conservation easements have been completed in Northwest Montana, and new ranchland easements are created every month across the state. In 2023, the US Fish and Wildlife Service proposed designating the headwaters of the Missouri River for 250,000 acres of conservation easement purchases, including more land in the Big Hole Valley. Important conservation is also taking place outside the eight landscapes we focused on. Purple and dark green now cover more and more of the maps, but that is a story for the next chronicle whenever the time is right. Just some closing thoughts are offered here.

Each region of Montana demonstrates a different path toward successful land conservation. Some efforts began opportunistically and evolved into a strategy, while others began as "large landscape conservation" visions. Important roles were played by national conservation groups, Montana land trusts, government land management agencies, tribes, corporations, and the Department of Defense. However, in all cases immense credit goes to Montana landowners who weighed their options, drew on their values, and left a legacy of good land. No single approach applies everywhere—many roads, one journey.

The Rocky Mountain Front: a Conservation Area, leveraged partnerships, tribal work

The Blackfoot: citizen-led conservation, grassroots Americana

The Greater Yellowstone: an ecoregion focused on the world's first national park

The Missoula Region: open space partnerships, corporate actions

Northwest Montana: forestland conservation, farms, Wild and Scenic Rivers

The Flathead Reservation: tribal land conservation, return of the Bison Range

The Helena Region: scenic landmarks, river corridors, military conservation

The Prairie: an NGO helps return bison to the Great Plains

Montana is a national leader in voluntary, compensating land conservation. People love this place and aren't shy about expressing it. Montanans are also practical people, so the successes explored in this book reflect pragmatic approaches that worked. New tools are still being invented and opportunities continue to grow, but challenges remain. Future work will take lots of capital, patience, and creativity as Montana expands in population and diversifies in ideologies. Nothing should be taken for granted, not even Montana's climate.

Conservation and Climate Change

by Dr. Steven Running, Nobel Prize winner

Montana's climate is warming about 0.5 degrees Fahrenheit per decade at the current trajectory. It has warmed by 2.7 degrees since 1950. However, with large year-to-year variability, there is no overall trend of increasing precipitation. In fact, because warming air evaporates surface water faster, the net consequence is that Montana is slowly becoming more arid—it has a more negative water balance. The state's climate is sliding toward the current Utah climate. In response to these trends, ecosystems, both plants and animals, will tend to move from their present locations to maintain required living conditions. The initial pattern in our mountains may simply be movement from hotter south and west aspects to cooler north and east aspects. Next may be upward movement in elevation since timberlines in mountains around the world are rising. Over centuries ecosystems will move northward in latitude. The USDA is already adjusting horticultural hardiness zones northward. Forward-looking conservation efforts need to plan to accommodate this regional movement. The Yellowstone to Yukon Conservation Initiative is a good example of a continental-scale conservation vision.

Ecosystems, like humans, can survive for decades in suboptimal and even stressed conditions. However, when wildfires and insect epidemics ultimately kill stressed forests, these ecosystems may not regenerate as forests but as more arid, open shrubland or grassland communities. So, land conservation cannot simply aim to return ecosystems to a past "natural" condition but must recognize new and changing bioclimatic limits.

With the world population at eight billion and another one to two billion on the way, conserving a natural landscape in Montana is an ongoing priority. Coastal residents in the Gulf States are enduring sea level rise and more powerful hurricanes. In the Southwest, groundwater is being depleted and rivers are being dewatered. When I watch summer temperatures in our southern states reach levels that are hazardous to human health, I can't help but think that more of their populations are going to opt for Montana. Climate migration has started and could accelerate rapidly as more people understand the trajectory of global warming consequences. I hope that efforts to conserve our natural landscapes will remain a priority for Montana right now and into the future.

Hope

Jane Goodall is a renowned champion for chimps and our other primate cousins. After nearly 70 years, she continues to work against habitat loss, poaching, the "pet chimp" trade, and other threats. Jane has no university training, but a profound empathy and intelligence. Her work in Africa is based on quietly observing chimps to discern who

they truly are and how we can help them thrive. There have been many successes and many setbacks, but she keeps going anyway. How does Jane do it? In *The Book of Hope*, she explains. "Hope is a human survival trait. It is important to take action and realize we can make a difference, then this will encourage others to act. We will realize we are not alone and our cumulative actions truly make a great difference. That is how we spread the light, and in turn, become ambassadors for hope."

Making Certain It Goes On

Richard Hugo came up with that phrase. He was a Montana poet and fisherman who left us with deep notions about living here. Dick hoped that our love of the state would inspire us to live good and meaningful lives despite the tough times we face. His poems about Philipsburg, Kicking Horse Reservoir, trout, and taverns are a meeting place between geography and the human heart. His phrase "making certain it goes on" seems to mean protecting what is beautiful and natural around you. Live the best version of yourself in the presence of an immense natural inheritance.

All of this begins with personal reflection about what counts. For many of us that is best done while riding a cherished horse, hiking a favorite trail, or sitting beside home water. This provides an encounter with beauty and our true self, the one that reminds us about kindness and whispers, "Do the right thing." Sacredness is a feeling that flows to us from all traditions to form a confluence where land conservation gets done.

Here are a few things you can do to help.

If you own land, talk to someone you trust about conservation easements. Take your time, reflect on your values, run the numbers, and decide whether it works for you.

If you know someone who owns land, connect them with a source of honest information.

If you live in town, become part of an open space conservation program. Volunteer and lend a hand. Vote in favor of well-conceived open space bonds.

Tell your congressional representatives you support national land conservation programs and the funding that makes them work.

Work toward creating a state income tax credit for the donation of conservation easements. This wonky-sounding change will supercharge land protection.

Join a land trust or other conservation NGO.

Support conservation easement laws.

Donate to groups you respect.

Cherish the land.

Cultivate hope.

Conserve.

FURTHER READING

The Book of Hope: A Survival Guide for Trying Times. 2021. Jane Goodall, with Douglas Abrams and Gail Hudson. Celadon Books.

Greater Yellowstone Climate Assessment. 2021. www.gyclimate.org.

Montana Climate Assessment. 2017. www.montanaclimate.org.

APPENDIX A

Conservation Contacts

NONGOVERNMENTAL ORGANIZATIONS (NGOS)

American Farmland Trust (AFT). Pacific Northwest Office. PO Box 5263, Bellingham, WA 98227. (205) 453-5226. www.farmland.org.

Bitter Root Land Trust (BLT). 170 South 2nd Street, Suite B, Hamilton, MT 59840. (406) 375-0956. www.bitterrootlandtrust.org.

Blackfoot Challenge. PO Box 103, 405 Main Street, Ovando, MT 59854. (406) 793-3900. www.blackfootchallenge.org.

Boone and Crockett Club. 250 Station Drive, Missoula, MT 59801. (406) 542-1888. www.boone-crockett.org.

The Conservation Fund (TCF). Missoula Office. (406) 541-8555. www.conservationfund.org.

Ducks Unlimited (DU). Missoula, MT. (406) 581-8971. www.ducks.org.

Five Valleys Land Trust (FVLT). 120 Hickory Street, Suite B, Missoula, MT 59801. (406) 549-0755. www.fvlt.org.

Flathead Land Trust (FLT). PO Box 1913, Kalispell, MT 59903. (406) 752-8293. www.flatheadlandtrust.org.

Gallatin Valley Land Trust (GVLT). PO Box 7021, Bozeman, MT 59771. (406) 587-8404. www.gvlt.org.

Greater Yellowstone Coalition (GYC). 215 South Wallace Avenue, Bozeman, MT 59715. (406) 586-1593. www.greateryellowstone.org.

Heart of the Rockies Initiative. www.heart-of-rockies.org.

Institute of the Rockies. 802 East Front Street, Missoula, MT 59802. (406) 829-9378.

Kaniksu Land Trust. 2504 Tradewinds Way 4B, Thompson Falls, MT 59873. (406) 827-0487. www.kaniksu.org.

Montana Association of Land Trusts (MALT). PO Box 892, Helena, MT 59624. (406) 200-8360. www.montanalandtrusts.org.

Montana Land Reliance (MLR). PO Box 355, Helena, MT. 59624 (406) 443-7027. www.mtlandreliance.org.

Mule Deer Foundation (MDF). 1785 East 1450 South, Suite 210, Clearfield, UT 84015. (801) 973-3940. www.muledeer.org.

National Wildlife Federation (NWF). 970 Wyoming Street, Missoula, MT 59801. (406) 721-6705. www.nwf.org.

The Nature Conservancy (TNC). 32 South Ewing, Helena, MT 59601. (406) 443-0303. www.nature.org.

Prickly Pear Land Trust (PPLT). 40 West Lawrence Street, Suite A, PO Box 892, Helena, MT 59624. www.pricklypearlt.org.

Rocky Mountain Elk Foundation (RMEF). www.rmef.org.

Trout Unlimited (TU). Chapters exist all over Montana. www.montanatu.org.

Trust for Public Land (TPL). 1007 East Main Street, Suite 300, Bozeman, MT 59715. (406) 522-7450. www.tpl.org.

Vital Ground Foundation. 30 Fort Missoula Road, Missoula, MT 59804. (406) 549-8650. www.vitalground.org.

The Wilderness Land Trust. PO Box 881, Helena, MT 59624. (406) 397-5340. www.wildernesslandtrust.org.

Wildlife Land Trust. Humane Society of the United States. 1255 23rd Street NW, Suite 450, Washington, DC 20037. www.humanesociety.org/wildlife-land-trust.

FEDERAL AGENCIES

National Park Service (NPS). www.nps.gov.

US Bureau of Reclamation. PO Box 30137, Billings, MT 59107. (406) 247-7296. www.usbr.gov.

US Fish and Wildlife Service (USFWS). Montana Office. 4052 Bridger Canyon Road, Bozeman, MT 59715. (406) 585-9010. www.fws.gov.

US Forest Service. Northern Region Headquarters. 26 Fort Missoula Road, Missoula, MT 59804. (406) 329-3511. www.fs.usda.gov.

TRIBES (SEVERAL RESERVATIONS INCLUDE MULTIPLE TRIBES)

Blackfeet Indian Reservation. Browning, MT. (406) 338-7521. www.blackfeetnation.com.

Crow Indian Reservation. Crow Agency, MT. (406) 679-4048. www.crow-nsn.gov.

Flathead Indian Reservation. Pablo, MT. (406) 675-2700. www.csktribes.org.

Fort Belknap Indian Reservation. Harlem, MT. (406) 353-2205. www.ftbelknap.org.

Fort Peck Indian Reservation. Poplar, MT. (406) 768-2300. www.fortpecktribes.org.

Northern Cheyenne Reservation. Lame Deer, MT. (406) 477-6284. www.cheyennenation.com.

Rocky Boy's Indian Reservation. Box Elder, MT. (406) 395-4476. https://www.bia.gov/regional-offices/rocky-mountain/rocky-boys-agency.

STATE AGENCIES

Montana Department of Transportation. 2701 Prospect Avenue, PO Box 201001, Helena, MT 59620. www.mdt.mt.gov.

Montana Fish, Wildlife and Parks. (FWP). 1420 East Sixth Avenue, PO Box 200701, Helena, MT 59620. (406) 444-2535. fwp.mt.gov.

CITY AND COUNTY AGENCIES

Bozeman City Parks. 600 Bridger Drive, Bozeman, MT 59715. (406) 582-2290. www.bozeman.net.

Gallatin County Open Lands Board. 311 West Main Street, Room 108, Bozeman, MT 59715. (406) 582-3130. www.gallatinmt.gov.

Great Falls City Parks and Trails Department. 1700 River Drive North, PO Box 5021, Great Falls, MT 59403. (406) 771-1265. www.greatfallsmt.net.

Helena Parks, Recreation and Open Lands. 316 North Park Avenue, Helena, MT 59601. (406) 447-8000. www.helenamt.gov.

Lewis and Clark County. Open Lands Program. 316 North Park Avenue, Room 230, Helena, MT 59623. www.lccountymt.gov.

Missoula County Parks, Trails, and Open Lands. 127 East Main Street, Second Floor, Missoula, MT 59802. (406) 258-4657. www.missoulacounty.us.

Missoula Parks and Recreation. 435 Ryman Street, Missoula, MT 59802. (406) 552-6000. www.ci.missoula.mt.us.

Index